# Frank Pantridge MC

# Frank Pantridge MC

# Japanese Prisoner of War and Inventor of the Portable Defibrillator

Cecil Lowry

Pen & Sword
**MILITARY**

First published in Great Britain in 2020 by
Pen & Sword Military
An imprint of
Pen & Sword Books Ltd
Yorkshire – Philadelphia

Copyright © Cecil Lowry 2020

ISBN 978 1 52677 733 1

The right of Cecil Lowry to be identified as Author of this work has been asserted by him in accordance with the Copyright, Designs and Patents Act 1988.

A CIP catalogue record for this book is
available from the British Library.

All rights reserved. No part of this book may be reproduced or transmitted in any form or by any means, electronic or mechanical including photocopying, recording or by any information storage and retrieval system, without permission from the Publisher in writing.

Typeset by Mac Style
Printed and bound in the UK by TJ International Ltd,
Padstow, Cornwall.

Pen & Sword Books Limited incorporates the imprints of Atlas, Archaeology, Aviation, Discovery, Family History, Fiction, History, Maritime, Military, Military Classics, Politics, Select, Transport, True Crime, Air World, Frontline Publishing, Leo Cooper, Remember When, Seaforth Publishing, The Praetorian Press, Wharncliffe Local History, Wharncliffe Transport, Wharncliffe True Crime and White Owl.

For a complete list of Pen & Sword titles please contact

**PEN & SWORD BOOKS LIMITED**
47 Church Street, Barnsley, South Yorkshire, S70 2AS, England
E-mail: enquiries@pen-and-sword.co.uk
Website: www.pen-and-sword.co.uk

Or

**PEN AND SWORD BOOKS**
1950 Lawrence Rd, Havertown, PA 19083, USA
E-mail: Uspen-and-sword@casematepublishers.com
Website: www.penandswordbooks.com

# Contents

| | | |
|---|---|---|
| *Acknowledgements* | | vi |
| *Timeline* | | viii |
| *Foreword* | | x |
| *Introduction* | | xii |
| **Chapter 1** | Early Years | 1 |
| **Chapter 2** | Enlistment | 6 |
| **Chapter 3** | The Japanese Invasion of Malaya | 18 |
| **Chapter 4** | Changi Prisoner of War Camp Singapore | 40 |
| **Chapter 5** | F Force and the Death Railway | 60 |
| **Chapter 6** | Homeward Bound | 88 |
| **Chapter 7** | Belfast and the RVH again | 94 |
| **Chapter 8** | Twilight Years | 115 |
| *Conclusion* | | 118 |
| *Appendix I:* | *James Frank Pantridge – Obituary by Professor Alun Evans* | 125 |
| *Appendix II:* | *People Who Knew the Real Frank Pantridge* | 134 |
| *Appendix III:* | *Famous People Whose Lives Have Been Saved by a Defibrillator* | 143 |
| *Bibliography* | | 149 |
| *Index* | | 151 |

# Acknowledgements

Writing a book is often a lonely and self-centred business; however, as most writers will tell you, there are always people in the background offering you advice and support and they need to be acknowledged.

Firstly, I must thank my long-suffering wife Liz, for putting up with my addiction to the war in the Far East and the long hours I have 'gone missing' whilst I worked on this manuscript.

Secondly, to Professor Alun Evans of Queen's University Belfast for his input and sound advice. Professor Evans was a colleague of Frank's in the Royal Victoria Hospital and he has been a tower of strength and information during my research.

Sincere thanks also to Frank's nephew Frank Pantridge junior, for giving me a detailed insight into the character of his uncle.

To Lady Mary Peters, probably one of the best-known sports personalities in Northern Ireland, having shot to fame in 1970 after winning gold medals in the Pentathlon and Shot at the Edinburgh Commonwealth Games. Two years later she won a gold medal in the Pentathlon at the 1972 Munich Olympic games. Mary took time out of her busy schedule to give me a huge amount of information about Frank whom she first met in a pub near Hillsborough. They became good friends and she was the driving force behind the commissioning of a portrait of him that now hangs in Queen's University, Belfast. She was also on the committee of the Pantridge Trust.

To Jim Dillon, ex-mayor of Lisburn who, despite a serious illness, spent time giving me a great deal of background into the man he knew so well.

To Dr Charles Wilson, one of Frank's protegees and one of the first doctors to go out on the streets of Belfast in Pantridge's first ambulance.

To my friend and FEPOW (Far East Prisoners of War) researcher Keith Andrews for obtaining Frank's Japanese record card and his release questionnaire.

To David Burns, Chief Executive of Lisburn and Castlereagh City Council, and to the Council for their help and assistance and for hosting the launch of this book.

To the Ulster Transport Museum, Belfast, for providing me access to the first portable defibrillator ambulance and allowing me to use a photograph from their collection.

Thanks also to Francis Pike, author of the excellent *Hirohito's War*, for permitting me to use several of his charts.

To my friends Paul Jarvis and Jean Johnston for doing an initial proof read on the manuscript.

To Conor Bradford, son of the deceased Roy Bradford, a former prominent Northern Ireland politician. Conor's father had embarked on a biography of Pantridge some years ago but unfortunately the two men had a disagreement and his book was never published. Conor gave me permission to use some of the draft material from the original manuscript.

To Rod Beattie and all at the Death Railway Museum in Kanchanaburi Thailand for their help. Rod Beattie is probably the world's leading expert on the Thai/Burma Railway.

Thanks also to all at Pen and Sword for having the faith to accept this book for publication, particularly Henry Wilson, Matt Jones and my editor Irene Moore.

# Timeline

| | |
|---|---|
| 1916 | Born in Hillsborough County Down |
| 1928–1934 | Friends School Lisburn |
| 1934–1939 | Queen's University Belfast |
| 1939 | Graduated Bachelor of Medicine (BM) |
| 1941–1942 | The Gordon Highlanders, Singapore |
| 1942 | The Fall of Singapore/Changi |
| 1943–1944 | The Thai/Burma Railway |
| 1944–1946 | Changi Singapore |
| 1945 | Repatriation – Belfast |
| 1946 | Graduated Medical Doctor (MD) |
| 1947 | University of Michigan |
| 1950 | Consultant Cardiologist – Royal Victoria Hospital, Belfast |
| 1966 | The portable defibrillator is born |
| 1969 | Officer of the Order of St John (OStJ) |
| 1974 | Fellow of the American College of Cardiology |
| 1982 | Retirement |
| 2004 | Death |

*Almost certainly, only one cardiologist has conducted pioneering work that saved my life and those of thousands of others. This distinction goes to Dr James Francis 'Frank' Pantridge, professor at Queen's University, Belfast, Northern Ireland. During a visit to Charlottesville, Virginia, I had a heart attack in 1972, which a mobile coronary care unit successfully treated with a Pantridge Portable Defibrillator. I owe my life to the invention of this former Japanese prisoner of war.*

Lyndon B. Johnson
President of the United States of America 1963–1969

# Foreword

## by Alun Evans, Professor of Epidemiology
## Queen's University Belfast

Frank Pantridge and his colleagues were housemen in the Royal Victoria Hospital, Belfast, when World War II was declared on 1 September 1939. That day, he and four others enlisted in the army, although they had to complete their medical contracts before they saw service several months later. Frank and three others were sent to the Far East, while Thomas Field, Frank's close friend, because he had served in the Territorial Army, was sent to North Africa. When Thomas learned that Frank was being sent to Malaya, he exclaimed: 'God help the Japanese!' but, considering what unfolded, 'God help Frank!' would have been more apposite.

The five went off to war clad in hand-stitched uniforms supplied by Thomas' father, who owned a clothing company. Only one of them, the son of the Professor of Medicine, did not return. Frank won his Military Cross during the operations in Johore and Singapore; the citation stated that Frank *'was absolutely cool under the heaviest fire and completely regardless of his own personal safety at all times'*. I asked Frank what had transpired that day and he said he was simply determined to make sure that all the injured received adequate pain relief.

After Frank's capture, his subsequent experiences on the Thai-Burma Railway were to shape the rest of his life in various ways. The diseases prisoners faced, in order of importance, were: malaria, dysentery, beriberi, tropical ulcer, and cholera. Frank's boiling of his drinking water to avoid dysentery predisposed him to beriberi, and this led to his subsequent experimental reproduction of the condition in pigs, which, incidentally, came courtesy of his family's farm.

Frank's witnessing of so much death probably didn't help him strike up and maintain relationships, which may explain why he never married. Also, having to fight so hard to preserve his own life gave him an abiding contempt for suicide. As Rosemary Evans, who edited Frank's biography *An Unquiet Life*, observed at his memorial service: 'He was driven to spread medical good through revulsion from the bad which afflicted a part of his life.'

Frank always stated, '*I never felt I had done my job properly unless the patient felt better for having seen me.*' As well as the excellent, innovative care he provided, his charm and quick-fire humour did a lot to make his patients feel better. Frank loved to quote the American bank robber, Willie Sutton, who when asked, 'Why do you always rob banks?', replied: 'Because that's where the money is.' In a way, this neatly encapsulates Frank's philosophy.

The military historian, Cecil Lowry, has fittingly concentrated on Frank's military career, and I commend his book to you very highly.

# Introduction

*As Frank Pantridge was suffering from beriberi, he was transferred in the back of a truck to Thanbaya hospital camp at the end of August 1944. Lying on his makeshift bed in a very bad state, he was still determined that he would not die despite his companion in the next bed doing his best to drag him down into the depths of despair. Each morning this man would call across to him: 'another bloody hopeless dawn'. The word hopeless was a feature of the man's vocabulary throughout the day and Frank did his best to ignore him, repeating to himself: 'I will not leave my bloody bones in Burma.'*

*Doctor Frank Pantridge was determined that he would survive the horrors of the previous two and a half years spent languishing in Japanese prisoner of war camps. He knew even then that he would return home to his native Northern Ireland and make a difference to the lives of his fellow countrymen.*

As an author, quite often an idea for a new book eludes you until suddenly, that eureka moment hits you between the eyes. The idea for this book arose from one of those moments, when I came across Frank Pantridge's autobiography *An Unquiet Life* when I was doing some research.

My specialism is the Second World War in the Far East and the Thailand/Burma railway where thousands of prisoners of war tragically lost their lives at the hands of a cruel Japanese captor. The Second World War in Asia remains for many people, a dim and distant war fought against a fanatical enemy who coveted Britain's far flung empire; not for me however, as my father spent the worst three and a half years of his life in a Japanese prisoner of war camp.

Towards the end of 1941, with war with Hitler's Nazis at its height, Japanese aggression in the east began to expand. Ten years previously, they had occupied Chinese Manchuria and had been at war with China since then. The Japanese now had their greedy eyes on British, French, Dutch and American possessions as part of what they called 'their expansive South East Asia Co Prosperity Sphere'.

Preoccupied with the war on his doorstep in Europe, the then British Prime Minister Winston Churchill took a calculated gamble that the Japanese would not dare to invade British possessions in the east, particularly British Malaya and Singapore. Singapore was the jewel in the British Empire's crown, a mighty fortress guarding the trade routes from the Andaman Sea to the South China and Java Seas. History now tells us that Churchill's gamble backfired and, in a matter of seventy days, from December 1941 to February 1942, both Malaya and Singapore had fallen to the Japanese invaders. Over 130,000 Allied soldiers were taken prisoners of war. Among those soldiers was my father Hugh Lowry, a private in the 2nd Battalion of the East Surrey Regiment, and Frank Pantridge, a doctor in the 2nd Battalion of the Gordon Highlanders. Three and a half years later, on 6 and 9 August 1945, both men's lives were saved when US President Truman authorised the dropping of the atomic bombs on the Japanese cities of Hiroshima and Nagasaki.

On liberation, Frank Pantridge and my father were walking skeletons, but with active minds and a burning desire to return to their families and make an impact on post war life.

I was born and educated in Downpatrick, County Down, Northern Ireland, and have always been interested in any prisoners of war of the Japanese who might have originated from my home country. From my small town alone, four men were taken prisoner when Singapore fell. My father was fortunate to survive his years in the Far East and return home at the end of the war when the Japanese surrendered. Had he, and Frank Pantridge, joined the thousands of men who lost their lives in that conflict and in the prisoner of war camps, I would not be writing this book today.

It can be problematic when writing about historical events, particularly relating to memories and memoirs of individuals, but, where possible, I have consulted primary sources of information. I acknowledge that some of the facts and figures may not be totally accurate, but this does not in any way detract from the horrors that Frank Pantridge and my father went through as prisoners of war. I have done my utmost to verify the statistics, but with events that took place almost eighty years ago, there will always be errors and I apologise in advance for any such errors. But biography, like history, is formed only of those tiny scraps of information that have been rescued from the ashes. The rest is speculation.

Born in Hillsborough just outside Belfast, Frank Pantridge graduated in medicine from Queen's University in 1939 just prior to the outbreak of the Second World War. When war broke out on 3 September 1939, the young, newly qualified doctor enlisted in the army, but he had to wait seven months before being posted for training to Beckett Park, Leeds. After completing his training, he was posted to Singapore where he joined the Gordon Highlanders as their Medical Officer. On 15 February 1942 when the island fell to the Japanese, Frank, along with my father and 137,000 Allied soldiers, became a prisoner of war.

After spending over a year in the infamous Changi prisoner of war camp, the young doctor was transported on 29 April 1943 up to the Thai-Burma Railway, later to be known as the Death Railway. As part of F Force, he was forced to march for 175 miles through dense jungles to the upper end of the railway near the Burmese border.

It was known from as early as the late nineteenth century that most sudden deaths occurred as a result of ventricular fibrillation (VF), a disturbance in the heart rhythm. When this happens normal heart activity becomes chaotic, blood circulation stops, and the person usually becomes brain dead within four minutes if this is not corrected. If Cardiopulmonary Resuscitation (CPR) is applied immediately, this can be delayed up to fifteen minutes. CPR was first used in 1891 by a Doctor Friedrich Maass who performed the first equivocally documented chest compression in humans.

Nine years later in 1900, at the University of Geneva, doctors discovered that VF could be reversed if an electric shock is applied to the exposed heart of an animal. The shock brings the heart to a momentary standstill and as the chaotic pattern of contractions is interrupted, the cardiac muscle cells have the chance to resume work in an orderly sequence again. In 1947 it was applied in Cleveland Ohio to the open heart of a 14-year-old boy with a successful result; his life was saved.

Later that year it was discovered that an electric shock could be applied to the chest after smearing with P85 jelly, with no surgical procedures necessary. In effect this marked the beginning of a revolution in the development of heart research.

The defibrillator was invented by an American, William B. Kouwenhoven in 1930. In 1947, another American, Carl Beck, tested the technique on a human patient and saved his life, resulting in a new phase in the medical field and, thereafter, the defibrillator passed through many stages of improvement.

On his return to Belfast, Frank reasoned that if the problem lay outside of hospitals, ventricular fibrillation should be applied where the event occurred, in the workplace, the home, in the street, or in an ambulance. Statistics proved that survival chances decreased by 10 per cent for every minute that passed after a cardiac arrest. He likened this to the eighteenth century when many soldiers injured in battle had died before they were able to be brought back to the medical units, usually stationed about three miles behind the front lines. He said:

*In 1792, Larrey, a young French army surgeon, noted the plight of the wounded. At that time French army regulations decreed that medical personnel should remain one league behind the battle area. It was usually twenty-four hours before the wounded reached the surgical depots and by that time, they were often either dying or dead. Larrey devised a light vehicle to take the surgeons and their equipment to the front line and thus revolutionised military surgery.*

In 1962, many hospitals in the UK opened specialist cardiac care units housing a defibrillator. During the 1960s, there were over 100,000 coronary attacks in the United Kingdom each year, with around 35,000 of them being sudden deaths in people under 70 years of age. Such thinking resulted in Frank looking for a way to take the large and bulky defibrillators in the hospitals out to the patients.

As a result of his research in the Royal Victoria Hospital Belfast (RVH), Frank produced the first portable defibrillator in 1965. Initially operated from a specially equipped ambulance and running off car batteries, it weighed in at around 70 kilos. This initial model gradually evolved into the small compact units so prevalent in workplaces around the world today. With his ambulance on the streets, Belfast at that time became the safest place in the United Kingdom to have a heart attack.

Frank Pantridge's autobiography, *An Unquiet Life*, went out of print many years ago and original copies are hard to find, but this biography retells his story, particularly from his perspective as a prisoner of war. It is a work of non-fiction. No names have been altered, characters created, or events distorted. If a comment appears in quote marks, it is verbatim from oral history, letter or another primary source from the people who knew him.

There are stories we stumble upon that are so rich in detail, imbued with meaning and resonant with epic events of the past, that they can consume a writer. This book is one of those stories.

Introduction    xvii

**A Personal Perspective**

My father joined the East Surrey Regiment in 1937, two years before the outbreak of war with Germany. Jobs were scarce for a 20-year-old in rural Ulster, and a career in the army seemed like a good option. A year later, after training at the Surreys' base in Kingston upon Thames, his regiment was posted to Shanghai, China on peacekeeping duties, setting sail on 1 September 1938.

Almost two years to the day later, with the Japanese threatening China, the Surreys were posted to Singapore, arriving on 1 September 1940. After six months training in Singapore they were posted to the small town of Alor Setar in north Malaya, near the border with Thailand. By now there was a real threat to British possessions in the east from a Japanese invasion, and the regiment was part of the 11th Indian Infantry Brigade, set to counter any such threat from overland.

Nine months later, on 7 December 1941, this threat turned into a reality, when my father and his colleagues braced themselves for a Japanese assault. For the next seventy days the Surreys were pushed back down the Malay Peninsula to Singapore before they were eventually all taken prisoner when the island fell on 15 February 1942.

For the next three and a half years my father suffered at the hands of cruel and brutal captors on the notorious Thai/Burma railway, his experiences mirroring those of the subject of this book.

I should make it clear from the outset, that whilst my father's experiences were similar to what Frank Pantridge went through as a prisoner of war, it's highly unlikely that they knew each other or that their paths crossed in the Far East. My father was in one of the early parties to leave Changi for the railway in October 1942, whilst Frank did not leave until April 1943. There is a good chance though, that their paths may have crossed back in Northern Ireland after the war. My father was heavily involved in the British Legion and carried the Northern Ireland British Legion standard at many official events, including the Festival of Remembrance at the Royal Albert Hall. With both men sadly long gone, there is little chance of ever finding out.

*Chapter 1*

# Early Years

James Frank Pantridge was one of three children born to Elizabeth and Robert Pantridge, in the small village of Hillsborough where the family owned a small farm. Hillsborough is 12 miles from Belfast, the capital of Northern Ireland and a mere 18 miles from my own birthplace in Downpatrick. As a small backwater of around 3,000 inhabitants, it is best known as being the nearest village to the infamous Maze prison where many members of the Irish Republican Army (IRA) and Loyalist prisoners were housed during the 'Troubles' in Northern Ireland of the 1970s and 1980s.

The Pantridge family farm overlooked Northern Ireland's Government House where the Anglo-Irish Agreement was signed by British Prime Minister Margaret Thatcher and Irish Taoiseach Garrett Fitzgerald on 15 November 1985, the treaty between the United Kingdom and Ireland that helped bring an end to the 'Troubles' in Northern Ireland. The Pantridge farm also overlooked the Royal County Down Corporation of Horse Breeders racecourse, the second oldest course in the UK.

Hillsborough, an attractive plantation village, was one of many set up by James VI during the early part of the seventeenth century when he saw the planting of Scottish and English settlers in Ireland as a way to control, anglicise and civilise Ulster. In 1690, William of Orange camped overnight in a fort near to the village, on his way to fight the infamous Battle of the Boyne.

At the age of five, the young Frank was sent away to a preparatory boarding school in Magheralin, some eight miles from the family farm. His father and mother were keen for their second son to get the best possible education and for him to enter one of the professions. The young Frank, however, hated his early years at boarding school and constantly ran away back home. He wrote later about his times at primary school, '*I was declared persona non grata.*' His parents then decided to enrol him into the small Downshire school in their home village where he received a good grounding in the three Rs (reading, writing and arithmetic), along with a very disciplined approach to learning.

The village doctor in Hillsborough, Dr Boyd, was a friend of the Pantridge family and often visited the farm on horseback. Boyd was trusted by everyone in the village and he had a great influence on the young Frank. Even at that young age, he knew that he wanted to be a doctor and follow in Boyd's footsteps.

He was also fascinated by anything military. At the Downshire School, he learnt that the Ulster Division fought in the Battle of the Somme and that 5,500 men had died in thirty-six hours. The entire male population of the small village of Dollingstown, near Mageralin, was wiped out. He loved the war memorial in the village and would often walk the mile and a half from his school to gaze at it and read the names of the men who had died in defence of their country.

Besides the military, Frank's other big interest as a child was horses, often riding bareback on an old mare on the farm, and falling off regularly about which he reflected later:

*Perhaps the numerous falls on my head may well have had something to do with my juvenile delinquency. I was wrongly accused of stealing apples from the vicarage. I greatly resented the false accusation. There was no justice in the world, I remember concluding.*

Whilst at the Downshire, the family were dealt a stunning blow when his father died in 1926; Frank was 10 at the time. At the age of 12, he left the Downshire school and was sent to Friends' School, Lisburn, a well-respected Quaker school, five miles from the family farm. At Friends he was an average student, not excelling in any subject, but he enjoyed sport, especially playing cricket for the Hillsborough cricket team whilst still in his early teens. '*Slip fielding was to be the extent of my sporting prowess*,' he wrote later.

At Friends he became aware for the first time of the sectarian divisions that dominated Ulster. One day, whilst waiting for a bus in Lisburn, he was punched in the face by an older Catholic boy simply because he was wearing a Friends blazer that signified him as a Protestant. On the bus his friends shouted to him that he 'should have killed the Catholic bastard'. Ironically, at the time, his parents sent him and his brother to Reilly's Trench, the nearby Catholic church. It seems that Frank was very tolerant of both sides of the 'divide' in Northern Ireland, his nephew stating that he remembers asking him just out of interest if such and such a person was a Catholic or a Protestant, and his reply was '*It wouldn't matter.*'

As he progressed through his school life, Frank became more interested in medicine, whilst his brother Herbert was groomed to take over the family farm. He discovered that for some mysterious reason, a pass in Latin was essential to get a place at medical school and he had to work hard to pass the Latin exam. He did well enough at school to get into Queen's University Belfast (QUB) to study medicine, entering that institution in September 1934 at the age of 18. For the first two years at QUB he studied chemistry, zoology, physics, anatomy, physiology and botany, before entering hospital and seeing patients for the first time whilst in year three.

During his second year at QUB, just before the Easter recess in 1935, he fell off his stool in the dissecting room. His lecturer immediately diagnosed diphtheria and he was taken by ambulance to an isolation ward in the RVH. They did not check his blood pressure, nor was he given an electrocardiogram, but he was simply administered a diphtheria anti-serum. Such a basic examination would have annoyed him immensely as it was discovered later that he had had heart block, an abnormal heart rhythm where the heart beats too slowly.

Whilst recovering, he developed an interest in cardiology after he had been taken to see a physician called Boyd-Campbell. After doing some research, he discovered that vomiting was an indication of cardio-vascular collapse and in association with heart block was almost invariably fatal, thus his interest in the workings of the heart was aroused.

In 1937, Frank entered the RVH in Belfast for the first time, other than as a patient. He was to embark on his junior doctor training and, although he did not know it at the time, it was to be the hospital where he was to spend most of his later professional life. The RVH, designed in 1899, was completed in 1903 to commemorate the Diamond Jubilee of Queen Victoria and was opened by King Edward VII that year. It was the first air-conditioned hospital in the world. The Royal, as it is still universally known, was one of the most modern hospitals in the UK at that time. When it was opened, it was still a voluntary hospital, funded mainly by Sir William Pirrie, chairman of the Harland and Wolff shipyard. H & W were one of the biggest employers in Belfast and it was the yard that built the legendary *Titanic*, launched in 1911.

A year later in 1938, Frank took his final exams at the City Hospital Belfast. One of his questions was to diagnose a problem with a 10-year-old child. The boy had water on the lung, which Frank diagnosed as a pleural effusion. His examiner disagreed and was scathing: 'If you cannot recognise pneumonia, you will never practise medicine,' he said. Whilst Frank was convinced that his diagnosis was right, he could not argue with an examiner. He waited until the examiner had gone before persuading the house physician to take a sample of fluid from the child and have it tested. Later he called at the examiner's house with a specimen of the 'non-existent' pleural fluid. The test proved he was right and he passed with honours.

Graduating from QUB in June 1939, he obtained his first post as a locum in a GP practice in Comber, a small town 10 miles southeast of Belfast. On his first day at the surgery he found himself covering for the GPs' three weeks holiday. With three imminent maternity cases on his hands, Frank was concerned. Whilst he had been taught the principles of pregnancy at medical school, he was terrified that he might have to deal with an actual birth. Fortunately, he was not involved in any of the three deliveries as the local midwife looked after everything, with the children born successfully without complications.

After finishing his three weeks locum in Comber, he obtained a post as house physician in the RVH on 1 August 1939.

## Chapter 2

# Enlistment

At that time, the RVH Belfast had a close link to the military going back many years. Quite a few medical staff from the hospital had joined the three services prior to the First World War. One particularly distinguished member of staff was Sir William McArthur, a houseman in the hospital when it first opened in 1903. McArthur had been awarded the Distinguished Service Order (DSO) for his actions in saving lives in Flanders during the First World War and he was a great speaker to the medical students. Frank recalled a talk where he said with a flourish, 'dysentery is caused by the filthy feet of faecal-feeding flies fouling food'. At that time the young student did not realise the significance of McArthur's talk, the subject of which he was later to find out first hand.

Another doctor from the RVH, John Alexander Sinton, had been awarded the Victoria Cross (VC) for his actions whilst serving with the Indian Army in Baghdad in 1916. Such illustrious men provided Frank with a burning desire to join the Royal Army Medical Corps (RAMC) and serve his country, a desire that was very soon to be fulfilled. It was a mere two days after Frank started his post at the RVH on 1 September 1939, that, along with several of his colleagues, he was listening intently to the radio when Prime Minister Neville Chamberlain declared war with Germany:

> *I am speaking to you from the cabinet room in 10 Downing Street. This morning the British Ambassador in Berlin handed the German government a final note stating that unless we heard from them by eleven o'clock, they were prepared at once to withdraw their troops*

*from Poland, a state of war would exist between us. I have to tell you now that no such undertaking has been received and that consequently this country is now at war with Germany.*

The following day, along with several of his fellow housemen, Frank headed for the army recruiting offices in Belfast to join up.

It was to be another seven frustrating months before he received his call-up papers, and in April 1940 he reported to Beckett Park near Leeds for training. There he found the drill sessions and route marches boring and was itching to do some important proper medical work and get to the heart of the action (no pun intended). '*I developed an ache in my right arm from saluting and returning salutes, and made it known that I would go anywhere to get away from that dreary depot,*' he said.

When his first proper posting eventually came through, it was to join a draft to Norway. At last he would see some action, but because of a lack of skiing experience he was taken off the party at the last moment, a decision that was fortuitous for him as none of that Norway draft survived the war.

A few days later Frank was issued with tropical kit and put on a draft for Singapore. Japan had been making threatening noises towards the British possessions of Malaya and Burma as she desperately wanted to expand her empire. The attractive rubber and tin resources of Malaya were deemed to be a prime target and Winston Churchill decided to bolster the country's defences by increasing the armed forces based there.

Since it was founded in 1819, the island of Singapore had been crucial for the defence of British possessions in the east. During the 1930s almost twenty-five per cent of the British Empire's trade passed through Singapore docks. Its huge naval base, built at a cost of £25 million over the previous fifteen years, lay at the heart of its defence strategy. To defend the naval base, heavy 15-inch naval guns were stationed at Johore battery, Changi, and at Buona Vista, to deal with battleships. Three of the guns were given an all-round traverse and subterranean magazines.

By 1940, with Allied forces routed in France, Holland and Belgium by the Nazis, Japan was casting its eyes towards the oil rich countries of south east Asia. In September 1940, they occupied French Indochina under an agreement with the puppet Vichy government in Paris, ramping up tensions in the British, Dutch and American possessions in the east. The situation became critical when the Americans imposed sanctions on Japan, freezing their assets, blocking trade between the two countries and holding back supplies of petroleum. Australia stopped supplying Japan with scrap metal and Britain blocked the supply of tin and rubber from Malaya. If the Japanese were to have any chance of completing their annexation of China, they would have to quickly find new supplies of raw materials. It now seemed very likely that the threat to Malaya could soon become a reality.

Nine months after he had joined up, Frank finally set off in February 1941 for the Far East. The first leg of his journey was overland down through France to Marseille on the Mediterranean coast rather than by sea. By now there was a serious threat to allied shipping from German U-boat patrols in the Bay of Biscay, so overland was a much safer option. At Marseille he boarded the SS *Strathmore*, a small cruise ship that had been converted into a troop ship, bound for Singapore.

The fleet headed across the Mediterranean, stopping briefly in Malta before entering the Suez Canal. After stops at Aden and Bombay the *Strathmore* eventually steamed through the Straits of Malacca and round the tip of Tanjung Piai, before arriving into Keppel Harbour, Singapore on 1 April 1941.

Frank's first impressions of Singapore were not good. The harbour was similar to many docklands around the world, with steaming cranes, compressors, ships belching acrid smoke and the constant rattle of chains at the dockside. He recalled:

*The atmosphere was of oppressive heat and humidity. As we disembarked, I felt as though I was enveloped by a hot, wet blanket. I was destined to feel like this for much of the next five years. I was in a steaming hothouse a few miles from the equator.*

As the new arrivals walked down the gangplank, a fleet of open lorries stood nose to tail waiting to take them to their barracks.

Whilst Frank's initial impression of Singapore was poor, it improved during the drive through the city. He was amazed by the sights and sounds, the colours of the tropical flowers in the parks, the washing hung out on bamboo poles from the windows of hundreds of shanty houses and the smell of fish drying on the pavements. It was bright and warm, the air smelt of fresh fruit and rickshaws flitted about the streets as the lorry lurched along.

Singapore, the City of the Tiger, was a lively cosmopolitan bustling city when he arrived in 1941, probably more cosmopolitan and energetic than even London, Calcutta or New York. The sea sparkled all around and there were patches of green everywhere. Sports grounds, golf courses, parks and gardens abounded, it certainly was the last resort of yesterday in the world of tomorrow.

After settling into his camp, Frank and his colleagues took the opportunity to enjoy the shops, clubs and bazaars that abounded. The New World, Happy World and Great World Clubs, were particularly popular places for the newly arrived troops and drinks were cheap. Some of the men took full advantage of the more specific 'hospitality' on offer in the plethora of bars and clubs, particularly in Bencoolin Street. Any thought of an impending war with the Japanese was pushed to the back of their minds as the troops lapped up the delights of this 'paradise in the east'.

Malaya was producing more than forty per cent of the world's rubber and thirty per cent of its tin, and both commodities were in high demand. With the war in Europe raging, the high-powered colonial industrialists in Singapore were leading a high life. With the dollars rolling in from sales, they wanted for nothing, including the best scotch whisky from Speyside and fresh oysters from Australia.

With plenty of money in their pockets, the newly arrived soldiers continued to enjoy the 'delights' of the many establishments in Singapore city. Such 'delights' often resulted in quite a few of them contracting venereal disease, mainly from the girls operating in the small seedy bars around Bencoolin Street. As a doctor responsible for the health and fitness of his men, Frank conducted a concerted campaign to try and eliminate VD, giving talks to the men outlining the nasty results of infection. He would say to the men, '*don't put your thing where any sensible man would not put a walking stick*'. He tried without success to get some of the red-light districts put out of bounds to the soldiers. It wasn't only the other ranks that took advantage of the 'hospitality' on offer, the officers also did so, but they tended to frequent the first-class hotels and clubs where the girls were reputed to be of a higher class.

No one in Singapore during the late 1930s however, seemed to take the Japanese threat seriously as the island was seen as an impregnable fortress. The Imperial Japanese Army were perceived to be useless fighters, incapable of firing a rifle, pilots wore glasses and could not fly at night. One of the colonels of a British infantry regiment in north Malaya at the time said, '*I hope that we are not getting too strong because if so, the Japanese may never attempt a landing.*' A year later a single Japanese tank column wiped out a third of his regiment.

The new Officer Commanding Malaya, Lieutenant General Arthur Percival, (known to the men as 'the chinless wonder') when asked about whether the defences of the island should be increased said, *'defences are bad for morale – for both troops and civilians'*, a statement that was later to come back and haunt him.

With the political situation deteriorating rapidly, training for the British troops in Malaya and Singapore was increased to an even higher tempo. At the age of 62, Air Chief Marshal Sir Robert Brooke-Popham was brought out of retirement to take over as Commander-in-Chief, Far East. Brooke-Popham was so entrenched in the Royal Air Force (RAF) that he claimed the obsolete Brewster Buffalo was, as he said at the time, *'more than a match for the Japanese Zero'*. In fact, the Zero was a much better aeroplane than either the Hurricane or the Spitfire and there were 'zero' of either of them in Singapore or Malaya.

The military command was convinced that any Japanese landing would come from the sea due to the difficulties of the terrain down the Malay Peninsula, even though such a threat overland had been flagged up some years previously. Rivalry between the services was also a problem in Malaya and Singapore. Twenty-two airfields had been built by the RAF, without consulting with the Army, who would have to defend them from any invading forces.

Defences around the coastline of the island were sparse. Lieutenant General Percival considered it bad for the morale of the Singaporeans, even though the Chief Engineer, Brigadier Ivan Simpson, had canvassed for months to have the fixed defences on the north coast of the island opposite Johore expanded. All three services had their own agendas and the leaders frequently fell out over matters of defence.

Three months after Frank's arrival in Singapore, Winston Churchill sent out Alfred Duff Cooper to mediate and to try and co-ordinate the commands under one War Council. Duff Cooper was appointed resident Cabinet Minister for Far Eastern Affairs, with the authority to set up a War Council. In a secret and personal letter to Churchill on 18 December, Duff Cooper was scathing in his contempt for senior colleagues on the island; he wrote:

> *Thomas* [Sir Shenton-Thomas, Governor of the Straits Settlement] *has found it impossible to adjust his mind to war conditions and was the mouthpiece of the last person he speaks to. There are no air-raid shelters, no trenches even, no tin hats or gas masks for the civilian population.*

His strongest attack was reserved for Thomas's decision that European women and children should not be evacuated from the threatened state of Perak, because a previous evacuation of white women from Penang had had a bad effect on the Asiatic population of Singapore:

> *It is the first time in the history of the British Empire when it has been our policy to evacuate the troops first and leave the women and children to the tender mercies of a particularly cruel Asiatic foe.*

Duff Cooper quickly fell out with Brooke Popham and Shenton Thomas, resulting in a chaotic chain of command. The RAF had only 158 aircraft, mostly obsolete, scattered around airfields over the Malay peninsula. Churchill wrote later:

> *The possibility of Singapore having no landward defences no more entered my mind than that of a battleship being launched without a bottom.*

Frank's first appointment in Singapore was to the military hospital at Tanglin, where his commander was Colonel Craven. On his first day, when Craven ordered him to count and record the empty beds in a hut, he replied tartly:

> *My only qualification is a degree in medicine, Colonel, perhaps a clerk would be better at simple arithmetic. I was then informed that disobedience of a direct order was mutiny, and the penalty for*

*mutiny was death. I survived the ultimate penalty but was given a punishment posting.*

Never afraid to speak his mind, this was the beginning of several 'run ins' Frank had with authority, and a few days later, Craven arranged for him to be posted to the 2nd Gordon Highlanders as their medical officer. The Gordons' commanding officer at the time was Lieutenant Colonel W.J. Graham, an authoritative figure.

The Gordons were stationed on the east of the Island at Selerang near Changi, and several other officers who had arrived in Singapore with Frank were also detailed to join the regiment. They included Captain Ludvig Gordon Farquhar from London, a fine architect who had a black belt in judo, and Second Lieutenant Robin Fletcher from Aberdeen who became the battalion adjutant. (Fletcher had been a teacher of foreign languages back in his native Scotland and his linguistic skills were to prove invaluable later in the PoW camps when his basic Japanese enabled him to converse with them.)

Second Lieutenant Geoffrey Hallowes from London managed to avoid being taken prisoner when he escaped from Singapore a matter of hours before the surrender on 15 February 1945 in a small dinghy. Along with Major Nicholson of the Royal Engineers, he succeeded in reaching Padang on Sumatra and thence on to Ceylon. Hallowes later married the famous French resistance fighter Odette Churchill, the first woman to be awarded the George Cross, on 20 August 1946. She was to become the most highly decorated woman, and most decorated spy of any gender, during the Second World War.

Second Lieutenant George Roberts from Edinburgh was a Scottish international rugby player. Tragically, he was to die at Kinsaiyok PoW camp in Thailand two and a half years later on 2 August 1943. Captain George Gray Grant from Huntly was the battalion transport officer and part of the famous Grants whisky family. Second Lieutenant Guillermo de Meir (known as Mo) was from Insch in Aberdeenshire, but his family originated from Mexico.

Mo was mentioned in dispatches after the war for his conduct as a PoW.

Second Lieutenant David Nunnely was from Devon. Second Lieutenant Duncan Campbell from Stonehaven was injured whilst mine laying in Johore on 20 January 1942. Second Lieutenant Derek Stewart from Rugby unfortunately died at the end of February 1942 when he was part of a detail sent by the Japanese into Johore to lift the mines laid by the Gordons a year earlier. Another officer was the godson of Queen Victoria. Frank, a boy from a small village in Northern Ireland, was now amongst an illustrious bunch of officers. '*The facilities at Selerang barracks were excellent, including a swimming pool and sports pitches. It was also adjacent to the beach where the men could enjoy swimming and beach sports,*' he recalled.

The regiment arose at 0400hrs and the men were on parade by 0600hrs each day. Usually training was completed by 1400hrs leaving them plenty of time for recreational pursuits.

Of course, despite the excellent facilities, Frank found life at Selerang boring. He craved some action. There was, however, the occasional incident for him to deal with. One day he was called to see a gunner with a dislocated shoulder and recounts the incident:

*One morning I was called to the medical reception station some four miles from Selerang to see a gunner with a dislocated shoulder. I got him to lie down, gave him a quarter gram of morphine and corrected the dislocation by placing my foot in his armpit and pulling on the arm. The colonel in charge was outraged. Surgical problems he said should be referred to a surgical specialist. I lost my temper and said what I thought as usual. The colonel became apoplectic at my temerity. There was seldom much rapport between the regular RAMC and those such as me with emergency commissions. He was a senior regular officer and I held a junior emergency commission. Once again, I found myself on an immediate court-martial charge, which, fortunately was quashed by the Assistant Director of Medical Services as medical officers were thin on the ground.*

Although the Gordons were a Highland regiment, most of the officers came from English public schools. There were four old Etonians, several from Ampleforth College, Downside and Stonyhurst and Frank from a small grammar school in Lisburn. Officers dressed for dinner, with much attention being attached to ceremony, as was normal for the British Army at the time, with every officer having a Chinese servant instead of a batman. Frank recalled: '*For the eighteen months after I reached Singapore life was unreal. Dunkirk, the capitulation of France, the Battle of Britain, the Balkans blitz, Rommel's success in North Africa, the German invasion of Russia in June 1941, all these events made little impact on the civilian population of Singapore.*'

Conditions in Singapore were so benign that he often wondered '*why the hell am I here. I had volunteered because there was war in Europe, and it appears that Malaya is unimportant. The priorities seemed to be support for the Russians and support for the Middle East.*' Such conditions of course were not to last.

One of Frank's friends in the Gordons was Lieutenant Ivan Lyon. Lyon was to go on to achieve fame during the coming war when, in February 1942 with Singapore doomed, he was charged with setting up an escape route across Dutch Sumatra called the 'Indragiri Line'. After the war, Lyon was awarded the MBE for his bravery in helping refugees to escape Singapore using this route. He himself managed to escape the clutches of the Japanese, eventually reaching Australia where he became determined to hit back at the Japanese. In September 1942 he led a daring operation called Jaywick, that managed to infiltrate Singapore Harbour and sink seven Japanese ships with limpet mines.

Frank celebrated his twenty-fifth birthday in Singapore on 3 October 1941 with celebrations going on into the small hours in the officer's mess.

Sections of the Gordons were training up on the Malaya mainland on the southern tip of Johore near Pengerang at the time and he spent some time travelling back and forward across the causeway. He recalled:

> *I spent much of my time commuting across the Johore Straits to Penerang where I was to meet some of the European rubber planters. They lead a lonely life and were glad of contact with any other Europeans. Some were whisky swillers, no doubt, but the majority were good characters and they were most hospitable towards the troops. A bottle of best Scotch cost one Straits dollar. One-pound sterling bought eight bottles. My memories of the weekend parties are hazy, but I recall that one young man fractured his skull by diving into the shallow end of a swimming pool at 4am on a Sunday morning.*

In early November 1942, Allied troops in Singapore and Malaya were put on a war footing. Whilst Frank was still in Singapore, my father, as a driver in the East Surreys, was stationed just north of the town of Jitra near the border with Thailand. The Surreys were part of the 11th Indian Division tasked with meeting any threat coming down the main road into Malaya.

In Johore, the Gordons worked tirelessly to lay anti-personnel mines on the beaches, build pillboxes and lay barbed wire to guard the vulnerable approach to the naval base. Unfortunately two of the men were killed and three others injured during this mine-laying exercise. The pillboxes later proved to be useless, as the materials used were found to be sub-standard, due to corruption by a British officer and a Chinese supplier. Frank recalled the boxes, '*an inspecting general was able to stick his walking stick through the so-called concrete. When the corruption was discovered, the officer was shot.'*

With the likelihood of a Japanese invasion looming even closer, the Gordons were moved back to Singapore, where they packed away all their personal possessions to be stored in the gymnasium of Selerang barracks. The regimental colours were deposited in a vault in the Hong Kong and Shanghai bank. The regiment was now on a full-scale war footing.

*Chapter 3*

# The Japanese Invasion of Malaya

Around 1400hrs on 6 December 1941, the crew of a Lockheed Hudson from No.1 Squadron Royal Australian Air Force (RAAF), flying on a reconnaissance mission out of Kota Bharu on the north-east coast of Malaya, spotted three large ships flying the Japanese flag, steaming west. The crew radioed back to their base, where Air Chief Marshal Sir Robert Brooke-Popham, Commander-in-Chief Far East, was advised of the sightings. Brooke-Popham ordered continued surveillance of the convoy, but no further action was taken.

The following morning an RAF Catalina flying boat was shot down over the Gulf of Siam, whilst almost simultaneously the Japanese attacked the American fleet in Pearl Harbor, Hawaii, bombed Wake Island, Midway Island, Luzon, Mindanao in the Philippines, Bataan Island and Kai Tak airfield in Hong Kong. The war with Japan had started.

Shortly after midnight on 8 December, a large Japanese force of over 12,000 troops with 400 vehicles and tanks, landed on the beaches of Singora and Pattani in east Thailand. By sunrise they were flooding south-westwards towards the border town of Betong near Kroh. Another Japanese regiment had landed further north and took over the Kra Isthmus, the narrowest part of the Malay Peninsula in southern Thailand, bordered to the west by the Andaman Sea and to the east by the Gulf of Thailand

Back in Singapore, a message came through to Lieutenant Colonel Stitt of the Gordons from Brooke-Popham, that read:

*Japan's action today gives the signal for the Empire's Naval, Army and Air Forces, and those of their allies, to go into action with a common aim and common ideals. We are ready. We have plenty of warning and our preparations are made and tested. We do not forget at this moment the years of patience and forbearance in which we have borne, with dignity and discipline, the petty insults and insolences inflicted on us by the Japanese in the Far East. We know that those things were only done because Japan thought she could take advantage of our supposed weakness. Now, when Japan has decided to put the matter to a sterner test, she will find out that she has made a grievous mistake.*

When General Percival called the Governor Sir Shenton Thomas to tell him the news. Thomas responded with words that would come back to haunt him, '*Well, I suppose you'll shove the little men off.*'

Back in the UK the following day, the *Daily Sketch* led with the headline:

**British Troops Are Mopping-Up the Malaya Invaders. Severe fighting has taken place in Malaya between British forces and Japanese invaders landed from the sea.**

It goes on to say:

*The Japanese are being mopped-up following air attacks on the warships and transports which landed them, and which were forced to retire. Fire was immediately opened by our troops, and later severe fighting developed on shore, particularly at the Kota Bahru aerodrome where our Indian troops are reported to have distinguished themselves. After daybreak air attacks were made against three aerodromes in northern Malaya but reports so far indicate that little damage has been done.*

Little did the people back in Britain know at that time what the next seventy days were to hold for their loved ones serving in the Far East.

The Royal Navy's two capital ships, HMS *Prince of Wales* and HMS *Repulse*, stationed in Singapore, along with four destroyers, left the naval base early on the morning of 8 December. They proceeded north through the South China Sea to engage Japanese troopships now known to be heading for the beaches of Thailand and Malaya.

Admiral Sir Tom Phillips, the commander of the convoy, had been warned by Brooke-Popham about the lack of air cover, but being old school navy, Phillips was arrogantly convinced that his ships would be able to inflict a heavy defeat on the Japanese invaders. Unfortunately, his stubbornness was to prove to be his nemesis, as both ships were sunk on 10 December by Japanese bombers operating out of French Indochina. Phillips, along with 1,870 of his crew, lost their lives that day when the two ships went down, effectively ending any chance of the Allies defending Malaya and Singapore. The loss of the two capital ships had a demoralising effect of the morale of the troops and the civilian population. Frank said at the time:

> *I saw the battleship* Prince of Wales *and the battle-cruiser* Repulse *accompanied by three destroyers sail from the naval base. There was no escorting aircraft carrier. Had there not been an incredible communication mess up, the RAF's Brewster Buffaloes could have reached the area before the ships were struck.*

As dawn broke on 11 December, the Japanese 5th Division crossed the border from Thailand into Malaya. By sunset they were engaging the forward positions near the town of Jitra, occupied by the 6th Indian Infantry Brigade, where my father was huddled under his cape sheltering from the heavy monsoon downpours.

Jitra soon became a major disaster for the 6th Indian Infantry Brigade, when a Japanese force of around 600 men drove them back, killing or capturing around 1,400 men, and taking great quantities of weapons and equipment. The Jitra position, that should have held out for three months, fell to the Japanese forces in a mere fifteen hours.

Whilst Frank and his colleagues in the Gordons waited back in Singapore, the Japanese continued to push the Allied forces back down the Malay Peninsula. For the next three weeks it was retreat after retreat for my father and his colleagues in the East Surreys.

With the Japanese now threatening the island of Singapore, the Gordons were moved to Gemas, 200 miles further north from Pengerang, where they joined two Australian battalions, the 2/26th from Queensland and the 2/30th from Sydney. The three battalions were tasked with holding a line across the main road, but without naval or air support they were soon ordered to retreat and set up roadblocks to try and halt the Japanese advance. Frank remembered the first roadblock in Gemas:

*An ambush was set up in the Gemas area of North Johore. In this the 2/30 Australian battalion commanded by 'Black' Jack Galleghan played a major part. We didn't have to wait long. The Japanese came on bicycles, followed by trucks and tanks. They were variously clad – some disguised as natives, others in bits of uniform. An already mined bridge was blown behind them. The Scots and Australians mounted machine guns on either side of the road and the killing area was covered by mortars. I recall arms and legs and other bits of bodies flying past me, though I was on a bank ten feet above the level of the road. The Gordons operated this way for a week or more. After retreating during the night, the first thing the troops did was dig in. The trenches dug in haste were relatively shallow, which didn't provide a lot of cover because the Japanese made good use of snipers strapped to the tops of trees. I had exchanged my service issue revolver for a tommy gun and found the bursts of fire directed at the tree tops proved effective.*

For a week the Gordons retreated down Johore, setting up roadblocks at dawn, holding them during the day and retreating again at nightfall. By now the men were exhausted, and Frank remembered the eerie feeling of marching night after night with virtually no sleep. '*It was like walking on air, I was never sure where my feet would land when I stepped forward.*'

As medical officer, he was always concerned about the welfare of his men. Despite his own personal discomfort, leaving the wounded behind was a constant worry, as he was well aware of the rumours that the Japanese had a reputation for beheading any wounded soldiers they came across. News had spread that the Japanese Imperial Guard had beheaded more than 200 Australian and Indian troops, run over their bodies with jeeps, saturated them with petrol and set them alight. Such rumours worried the retreating Gordons and Frank tried his best to move the wounded back to safety as quickly as possible. He was determined that those who could not be moved, would never suffer at the hands of the Japanese. The inference here is that they might even have put badly wounded men out of their misery, but he does not clarify this in his account of the retreat.

In Johore, the Gordons made contact with the newly arrived British 18th Division advance party, diverted from the Middle East. Frank recalled meeting the early arrivals of the 18th Division:

*They had been at sea for some considerable time. In the Malayan jungle they were quite lost and, it must be said, completely ineffective. They might as well have stayed at sea. They found the rubber plantations and secondary jungle a nightmare, an unknown world of snakes, insects, particularly leeches, and nocturnal noises.*

As they progressed down through Johore in pursuit of the retreating Allied troops, the Japanese were surprised at the lack of organised defence positions. A Japanese general is quoted as saying after the war, '*If strong defences had been vigorously defended in Johore, we might never have captured Singapore.*'

By 26 January the Gordons were dug in near a rubber plantation on the Ayer Hitam Road, a mere 50 miles north of Singapore island. They were attacked several times that day, with six men killed and ten wounded, keeping Frank very busy. For the next two days the regiment made a strategic withdrawal down through Johore towards Singapore. On reaching the Johore side of the causeway linking it with Singapore island, the Gordons, as part of the Australian 22nd Brigade, were tasked with holding a four-mile perimeter to allow the 3rd Indian Corps to pass through.

General Percival was well aware that by now the Japanese were massing their forces just across the straits, and that Singapore island itself was under huge threat. He was still convinced that any landings would take place on the north-east coast of the island, whilst General Wavell and Brigadier Simpson, the chief engineer, were convinced that the north-west coast was a more likely landing area. Percival overrode them and ordered Simpson to move all matériel to the north-east. Fortunately, the Japanese were now confining their efforts to sporadic air attacks on the island's airfields, harbour, city and military installations.

Due to the severity of the threat, the families of the Gordons still living in Singapore were now told to prepare to be evacuated and move their possessions to the Selerang gymnasium. On Thursday 30 January most of the families embarked on the *Duchess of Bedford*, sailing at 1800hrs. On board was Mrs Dorothy Brewster, wife of Colour Sergeant Bill Brewster and her one-year-old son. Mrs Brewster was originally from my home town of Downpatrick and in later years the Brewsters became good friends of my mother and father.

As the *Duchess of Bedford* steamed out of Keppel Harbour, Japanese bombs were dropping all around her. Fortunately, the ship was not hit and it arrived safely into Liverpool on 2 April some two months later. The *Duchess* was dubbed *'the most bombed ship still afloat'*.

Captain Ivan Lyon's wife Gabrielle and their five-month-old son also managed to get away from Singapore on a ship bound for Australia. Unfortunately, two months later in April 1942, en route to India from Australia, Mrs Lyon and the child were captured by the Germans when their ship was attacked in the Indian ocean. They were handed over to the Japanese by the Germans and were interned for the rest of the war.[1]

On New Year's Eve 1941, the Gordons, along with 30,000 other allied troops were moved back across the causeway to Singapore island. As they trudged across, they were accompanied by the two remaining pipers of their fellow Scots, the Argyll and Sutherland Highlanders. Accounts vary about the tunes that the Argylls played that day, from 'A Hundred Pipers', 'Cock O'the North','Hielan Laddie', 'Jenney's Black E'en' or 'Blue Bonnets O'er the Border.'

'*It was the beginning of the end for us, and, I might add, it proved to be the end of British influence in Asia*,' said Frank, as he trudged across carrying his medical kit.

A few hours later, the causeway was blown up by the Royal Engineers using naval depth charges. All road, rail and water links with the mainland were now severed, including the pipeline that brought fresh water to the island – Singapore island was isolated.

After crossing the causeway, the Gordons were given twenty-four hours rest at Birdwood Camp. A new order came through from General Wavell, an order that caused much amusement among the

---

1. Ivan Lyon was killed by the Japanese in July 1945, a mere month before the end of the war. He had led two very successful raids on Japanese shipping in Singapore Harbour from Australia before being caught trying to get away after the second one on 15 October 1944.

battle-hardened Scots. It said: *'Stand firm men. We are holding the Germans in the Middle East and you have got to stand here and fight and not turn back.'*

Frank wondered what he meant when he said, 'not turn back' as he knew if they did turn back, they would all have ended up in the sea. Their respite at Birdwood camp was disrupted regularly by Japanese air attacks causing several casualties as Frank recalled:

*I remember sleeping some of the time, only vaguely aware that the ground seemed to heave several inches each time one of the 15-inch guns was fired. Somehow, they had been swivelled round to point towards Johore beyond the causeway, but the armour piercing shells simply plunged harmlessly into the Johore hillsides.*

After their brief rest, if you could call it that, the Gordons were sent to guard the artillery installations protecting the vitally important naval base. This proved futile however, when the following day the base was evacuated with the staff all 'abandoning ship' and sailing for Ceylon. It was heart-breaking for the troops, who had fought so valiantly up country, to see the pathetic scene of an empty naval base, supposedly the pride of the British Empire. It was a base that had cost the British taxpayer millions of pounds to construct, and it was still packed with supplies waiting for the arrival of the invading Japanese forces.

Frank continued to deal with a wide range of casualties in the Gordons' regimental aid post located in a deserted house. Many of the injuries he had to deal with were horrible and some were bizarre. '*One Indian officer came in with a hand on either side of his head. He said that his head was about to fall off. He was right, I found a massive gash from a Japanese sword had severed all the muscles at the back of his neck down to his spine*,' he recalled.

On 5 February, five ships carrying the long-awaited reinforcements of the British 18th Division second wave, arrived into Keppel Harbour much to the relief of the troops already under heavy fire.

Led by Major General Beckwith-Smith, the troops of 54 and 55 Brigade landed along with a light tank squadron from India – the only tanks to reach Malaya during the entire campaign – and a number of Hurricane fighters. Their arrival lifted the morale of Frank and his fellow Gordons, but it was soon discovered that the newly arrived troops proved to be badly trained and had little idea about fighting a war in such a hostile environment. Frank had already discovered this several weeks earlier when he had met the advance party of the 18th up in Johore.

Most of Singapore was now covered by a thick haze of black, acrid smoke. Grey dust covered everything as the Japanese laid down fierce bombardments all over the island. The empty naval base was heavily bombed and when the large oil tanks were hit, thick black smoke blotted out the sun. A fine mist of black oil coated the troops from head to foot. It was the most intensive air and artillery barrage that most of them had ever seen.

As mentioned earlier, it had always been assumed that the Japanese would never attack the island of Singapore down the Malay Peninsula, particularly during the monsoon season, but such an erroneous assumption was now coming home to roost. The big guns on Singapore Island had been installed on the south coast to combat any invasion from the sea and the northern coastline had been neglected. For many years it was thought that these guns could not be turned around to face a threat from the land, however this was a fallacy. The guns were capable of traversing through 180 degrees to provide considerable firepower against any invasion across the straits of Johore. The only problem was that the gunners had only armour-piercing shells available to them.

As darkness fell on 8 February 1942, under a heavy covering barrage from guns on Johore, nearly 13,000 troops of the 5th and 18th Imperial Japanese divisions hit the beaches near to the Choa Chu

Kang and Ama Keng villages. By daybreak the next day they were attacking Tengah aerodrome, firmly establishing themselves on the island.

The morale of the defending forces, including the Gordons, had by now reached rock bottom. The city and the docks were now under constant bombardment and the outlook was bleak. The following day, in an attempt to raise morale, General Wavell sent another letter to the senior officers in Singapore penned in his usual inspirational language:

*We can name the nations who started the fire, Germany and Japan, stupid ill-bred children who have never been properly brought up or learnt good international manners. Their silly little girl friend Italy joined them in hopes of some cheap fun, and now finds herself being taken for a ride with two really bad boys. Japan has gained some initial successes, the successes that the murderer, the thief and the cheat can always gain over the honest decent citizen, until police have time to take steps to vindicate law and order and fair dealing.*

*It is certain that our troops in Singapore outnumber the Japanese troops who have crossed the Straits. We must destroy them. Our whole fighting reputation is at stake and the honour of the British Empire. The Americans have held out in the Bataan Peninsula against far heavier odds. The Russians are turning back the packed strength of the Germans. The Chinese, with almost complete lack of modern equipment, have held the Japanese for about four and a half years. It will be disgraceful if we cannot hold our much-boasted Fortress of Singapore against inferior forces. There must be no thought of sparing the troops or civil population. No mercy must be shown in any shape or form. Commanders and Senior Officers must lead their troops and, if necessary, die with them. There must be no thought of surrender and every unit must fight to the end and in close contact with the enemy.*

> *Please see that the above is brought to the notice of all Senior Officers and through them to all troops. I look to you and your men to fight to the end and prove that the fighting spirit that won our Empire still exists to defend it.*
>
> *Signed General Wavell. Commander in Chief South Western Pacific, 10 February 1942.*

The following day, Wavell flew out of Singapore bound for Java leaving the besieged troops to their fate. Some historians have accused him of running away, but he was ordered by Churchill to leave as his expertise was required for the defence of Burma and India.

On the evening of 11 February, the Gordons were moved to Tyersall Park, joining the Australian 22 and 44 Indian Brigades to form an area facing west along Reformatory Road. Their positions were attacked the following morning, but the two brigades were able to hold their lines under heavy air bombardment.

On the other side of the Island the Japanese were closing in on the city of Singapore itself. Casualties were mounting by the hour and the Alexandra Military Hospital came under attack. A friend of Frank's, Captain Smiley who had been in the same year in QUB, was one of the medical officers in the hospital at the time. Japanese mortar bombs began landing just to the rear of the hospital buildings, damaging the chapel and some other outbuildings and by early afternoon the shelling had increased. The hospital was now bursting at the seams, with around 800 patients and new casualties arriving by lorry load every hour; it was in total chaos. Patients were crammed into wards with many left lying on the floor or in camp beds lodged in every nook and cranny of the hospital, even the dining room had been converted into a temporary ward.

Just after 0800hrs on the 14th, heavy bombs started to land all around the hospital and by lunchtime Japanese soldiers were advancing across the hospital grounds. They soon entered the hospital and went berserk bayoneting surgeons, nurses and patients, some of them lying in the operating theatres.

What was to happen in the Alexandra Hospital over the next forty-eight hours was to go down in the annals as one of the worst massacres of the Second World War. The Japanese troops proceeded to murder in cold blood many of the patients and staff. The situation for the hospital now looked bleak. Lieutenant Colonel Craven, commanding officer of the hospital was discussing the gravity of the situation with his officer in charge of Radiography, Major Bull, in an upstairs office when a bullet came in through the open window causing them to throw themselves to the floor. The colonel decided to go downstairs to see for himself what was going on, but before he could leave the room, all hell broke out below, with explosions, gunfire and screaming echoing through the corridors.

The colonel decided to sit tight for a while, and after about half an hour, when things began to quieten down, he went down to the ground floor with his aides to assess the situation. The scene that greeted him was one of utter carnage. More than fifty patients and staff had been shot or bayoneted and many others had been taken away by the Japanese soldiers.

Some of the patients lying on the floor in the dining room saw, through the open doors, soldiers of the allied 44th Indian Brigade moving swiftly along the corridor from the medical wards, closely followed by a cluster of Japanese troops. Why the Punjabis were fighting in the hospital no one knows, but some people believe that the presence of armed troops in and around the hospital inflamed the situation causing the Japanese to react as they did.

The Japanese assault on the hospital was undertaken by a company of about 175 men. They were dressed in full combat kit, consisting of green tropical uniforms, steel helmets, rifles, bayonets and machine guns. They were heavily camouflaged with small branches, twigs and leaves stuck all over their bodies and were dirty and smelly. They also gave the impression to the patients of being either drunk or high on drugs. One group came in from the adjacent railway line and quickly entered the hospital through the main entrance. A second group came in through the rear entrance and the patients' dining room; a third entered through the operating room windows and through the surgical wards. Staff in the laboratory saw the first group coming towards the hospital; they tried to escape towards the main building but were tragically cut down by a burst of machine-gun fire.

Two of the second group of Japanese soldiers who had pursued the retreating Punjabi soldiers entered the temporary ward in the dining room where over 100 patients were lying on temporary beds. One of the two started beating the helpless patients with a brush whilst the other completely humiliated another by urinating on him. The rest of the soldiers in this group stormed into the room and began shooting and bayoneting many of the helpless patients. Among those killed in this early orgy of violence was Padre Smith of the Gordons, a very good friend of Frank's, along with Lance Corporal Simpson, Captain Harry Smith and Private Andrew Annandan, other soldiers that he knew well. With the hospital in a complete state of disarray and panic, many tried to escape from the marauding assassins, some with greater degrees of success than others. Those who tried to escape through the main entrance were simply mown down by rifle and machine-gun fire. Others met their fate in the rear corridor where they were bayoneted or shot before they had a chance to leave the building.

This group of Japanese soldiers seemed to have completed the worst atrocities. They entered the medical wards bayoneting patients and heavily beating up others. They forced the entire medical staff and wounded who could walk into the corridor, where they left them standing in sheer terror.

Whilst all this was going on, the surgeons carried on with the operations, firstly in the theatres, but when this proved too dangerous because of stray bullets, in the corridors, only moving back into the theatres when the firing stopped. Operations proceeded under the most stressful of situations and the sheer dedication and bravery of the medical staff was of the highest order.

The staff became aware that the Japanese were now in the corridors outside the operating theatres and Captain Dr Smiley moved over to the doorway and indicated the red cross clearly displayed on his armband. He motioned to the Japanese soldiers to come into the theatre to show them that they were not at any risk and that it was indeed simply an operating theatre. Smiley told the medical team to stand quietly in the centre of the room with their arms in the air leaving the patient on the operating table. The Japanese ordered the group out into the corridor and as they moved down the passage they were attacked viciously with bayonets and rifle butts. The captain took a bayonet thrust to his chest, but fortunately he had his metal cigarette case in his pocket and the bayonet was deflected. Captain Smiley managed to survive the frenzied attack by deflecting the bayonet away from his vital organs, although he was badly wounded in the groin, arm and hand. Frank's friend, Captain Smiley was awarded the Military Cross (MC) after the war for his actions in the hospital that day.

Whilst Smiley and Sutton lay motionless on the floor, they became aware of a large group of patients and medical staff running along the corridor with their arms raised. The Japanese were kicking and beating them with rifle butts as they stumbled along. Other soldiers entered several of the surgical wards and began to systematically beat up the helpless patients, many with fractures and in plaster casts. They took great pleasure in twisting and pulling at the slings holding some of the men's broken legs in the air.

Half an hour after the initial assault on the hospital around fifty staff and patients were dead, with many more lying around in great pain from their wounds. By 1530hrs over 200 patients and staff were roped together in groups of eight and forced to walk to a piece of open ground about 100 yards from the main buildings. A second group of around 60 (mainly officers) were taken out to a separate piece of land and roped together. Many of those in the first group were walking wounded with arms in plaster, most were in bare feet wearing only pyjamas. Those who had difficulty walking were supported by their comrades and if they fell, they were brutally bayoneted.

Those in this large group were then marched alongside the main railway embankment, past the Normanton fuel oil tanks which were on fire and belching huge clouds of black smoke into the hot and humid air. As they stumbled along, they also had to endure heavy shell fire from their own artillery causing them to dive for cover in the undergrowth on several occasions. At one point the party was told to stop and rest. They then had all valuables such as rings, watches, pens etc. taken from them and were again subjected to beatings with rifle butts and fists. One man whose arm was in plaster had it forced behind his back and re-broken – their brutality knew no bounds. By now their rope bindings had tightened in the heat causing considerable pain and some people's hands began to turn blue.

The building into which they were now forced was part of the old hospital's sisters' quarters. It was a red brick two-storey house raised above the ground on piles, with a block of outbuildings surrounding a small courtyard. The group were forced into three very small rooms, each with double doors opening directly onto the courtyard. They were crammed so tightly into the rooms that it was impossible for everyone to sit down at the same time. The doors were secured with wooden poles and the windows were nailed closed and shuttered with wood, preventing any daylight entering. There was no ventilation and within minutes the heat became unbearable for the prisoners. Their hands were still tied but they eventually managed to untie each other enabling them to raise their hands above their heads to allow more space. They were crammed in tighter than battery hens, had to suffer the indignity of relieving themselves on each other: of course, the smell soon became appalling.

The Japanese made no effort to provide food or drink and many of the men had not eaten or drunk anything for many hours. The conditions under which they were held were so inhumane that many of the men became mentally unstable. Many began to shout and scream at the tops of their voices appealing for food and drink. Others just slumped on the floor giving up the will to live.

By now it was late evening and for the rest of the night they endured the most difficult of situations. One of the officers said that food and water had been promised by 0600hrs the next morning, but the screaming and shouting went on all night. At one point a voice was heard from outside the rooms speaking in perfect English: '*If you keep quiet, I will try and get you back to the hospital tomorrow.*'

As dawn broke on the 15th, several of the men had passed away – at least seven in the central room alone. As the morning wore on with no sign of any food or water, desperation began to set in amongst the prisoners. Around 1100hrs a Japanese officer opened the door of one of the rooms and announced in broken English: '*We are taking you behind the lines – you will get water on the way.*' A few minutes later prisoners were taken out in pairs by the guards. Everyone assumed that they were being taken out for a drink until they heard screams and shouts in English from nearby. '*Oh my god – mother – don't – help me.*' It dawned on the remaining prisoners that that they were being executed; this was confirmed when a Japanese soldier was seen returning wiping blood from his bayonet with a large piece of cloth.

As the guards worked their way through the rooms removing prisoners in twos, the men in the last room became very distressed. Several tried to commit suicide – one by cutting his wrists and another by hanging himself. By mid-afternoon more than half of the men had been taken from this last room to meet their fate.

As all this was going on, heavy fighting was raging all around the buildings and a shell landing nearby blew open the doors and windows, showering the remaining prisoners with dust and rubble. Several were injured by the flying debris but those left could make a dash for freedom – some through the broken door and others through the windows. A number managed to get away but most of them were killed by a burst of machine-gun fire from close range.

Back at the hospital the cleaning up operation had begun, hampered by large numbers of very determined Japanese looters resulting in some unpleasant scenes. Many of the patients still in possession of their watches, rings, pens etc. were robbed and much of the hospital's food was stolen.

Lieutenant Moore, who was in the hospital at the time, recalls a Japanese senior officer entering after the massacre:

*He expressed his regret at what had happened and assured the staff that they had nothing further to fear. He also told the officer commanding the hospital that he was to be regarded as a direct representative of the Emperor and that no higher honour could be paid to the hospital.*

Six Gordon Highlanders died in the Alexandra Hospital massacre, including Captain Henry Smith, the chaplain who was bayoneted. Smith was in the hospital doing pastoral work when the Japanese entered.

That same day Churchill sent a terse signal to General Wavell:

*There must be no thought of saving the troops or sparing the population. The battle must be fought to the bitter end at all costs. Commanders and senior officers should die with their troops. The honour of the British Empire and the British Army is at stake. I rely on you to show no mercy or weakness in any form. With the Russians fighting as they are and the Americans so stubborn at Luzon the whole reputation of our country and our race is involved.*

Sitting in his comfortable office back in London, Winston Churchill had no idea what Frank, my father and the rest of the dispirited troops in Singapore were going through as the Japanese pushed towards the city.

The Gordons were stationed on Bukit Timah road when an order came through for them to advance towards the oncoming Japanese. This order was almost immediately cancelled, and they retreated once again to Holland Road, where they were again subjected to air attacks and mortar bombardment.

Brigadier Simpson, General Percival's Chief Engineer, proposed that a 'scorched earth' policy in Singapore should be adopted, but the Governor Sir Shenton Thomas was totally against such a policy. He said, 'the destruction of around forty large Chinese engineering works would be bad for morale.' As history will tell us, however, these works would soon be handed over to the Japanese completely intact.

During the final four days of the battle for Singapore, 17 men from Frank's regiment had been killed and 21 wounded, its worst period during the war. In total 62 men had been killed and 79 wounded during the seventy days of the conflict.

When the order to surrender came at 0800hrs on 15 February, the proud Gordons were reluctant to give up their weapons, but with Singapore doomed, they were forced to obey. There was to be no miraculous escape for the Singapore garrison, as there had been at Dunkirk almost two years earlier. Twice as many Allied troops were to fall into Japanese hands as fell into German hands at Dunkirk.

When Frank heard the news of the surrender he was astounded: '*I found the news unbelievable. I thought that it was to be a last man last round situation. Malaya and Singapore had been abandoned.*'

General Percival sent a letter to his commanders on that fateful day informing them of the capitulation:

*It has been necessary to give up the struggle, but I want the reason explained to all ranks. The forward Troops continue to hold their ground, but the essentials of war have run short. In a few days we will have neither petrol nor food. Many types of ammunition are short, and the water supply upon which the vast civilian population and many of the fighting troops are dependent, threatens to fail.*

*This situation has been brought about partly by being driven off our dumps and partly by hostile air and artillery action. Without these sinews of war, we cannot fight on. I thank all ranks for their efforts throughout this campaign.*

The fall of Singapore ended Great Britain's domination of East Asia and was the greatest defeat inflicted by an Asian army since Genghis Khan entered Vienna, more than seven centuries earlier.

Back in London Winston Churchill addressed parliament:

*I have not become the King's first Minister in order to preside over the liquidation of the British empire. We have so many men in Singapore, so many men. They should have done better. The Japanese moved quicker and ate less than our men,'* he lamented, rather bizarrely.

It is difficult to understand why Churchill deluded himself with the thought that Singapore could be held, considering the lack of available naval or air power and the chaos that reigned throughout the military command. Once Malaya fell, the island in isolation was undefendable. Some historians have said that Churchill had misjudged the situation in Malaya and Singapore, however, others have asserted that he deliberately turned a blind eye to the colonies, with Europe his main priority. Others have said that he sacrificed them in order to get the Americans into the war. I will leave the reader to make their own judgements on such assertions.

This controversy raged throughout the House of Commons back in London, with strident calls for a public enquiry. Churchill however opposed it and he said:

*I do not at all wonder that requests should be made for an inquiry by a Royal Commission, not only into what took place upon the spot in the agony of Singapore but into all the arrangements which had been made beforehand. I am convinced, however, that this would not be good for our country, and that it would hamper the prosecution of the war.*

Such an enquiry was never held, and whilst many files dating to that time have since come into the public domain, to this day there has never been an enquiry into the chaos regarding the fall of Singapore. But just as no single factor can be blamed for its fall, no one man can really be blamed either. According to Frank though, the lack of aircraft in Malaya and Singapore was one of the main contributors to the defeat: 'If some of the 336 aircraft promised had been available to the RAF, the scales might have been tipped in favour of the British. It was said that just a few of the planes sent by Churchill to Russia in 1941 could have saved Malaya,' he asserted after the war.

An article in *The Economist* at the time stated:

*Now the accidents of war have produced such a catalogue of catastrophes that the Prime Minister has to face something approaching a political crisis. This was not an accident such as might be encountered in war, but a disaster that occurred for a number of objective reasons; incapacity and poor training of the British troops, lack of resourcefulness and initiative on the part of the officers, poor strategy, inefficient administration, and indifference of the local population. The faults seem to fall into two categories: the errors and mistakes of the civilian administration and the ineptitudes of the military.*

In their planning for their assault on Malaya and Singapore, the Japanese took advantage of vital information that fell into their hands when the cargo steamer SS *Automedan* was captured by the Germans on 11 November 1940 just north-west of Sumatra. Documents disclosing that there was no British fleet available to help Singapore, found by the Germans in a chest on the ship, were passed on to the Japanese.

After the war it was discovered that the Japanese invaders were outnumbered three to one by the defending troops, a fact clarified by their General Officer Commanding Lieutenant General Tomoyuki Yamashita in his diary:

*My attack on Singapore was a bluff: a bluff that worked. I had 30,000 men and was outnumbered more than three to one. I knew that if I had to fight long for Singapore I would be beaten. That is why the surrender had to be at once. I was very frightened all the time that the British would discover our numerical weakness and lack of supplies and force me into disastrous street fighting.*

Frank Pantridge and my father were now just two out of around 130,000 Allied soldiers scattered around Singapore. Neither of them had any idea what was going to happen to them as they slept fitfully under their groundsheets waiting for further orders. For weeks they had lived with the deafening crash of gunfire, along with the sight and sounds of a battle for their lives, but now they were soon to be prisoners of war.

A deathly quiet hung over the city of Singapore, the smell of defeat and utter desolation was everywhere. The sight of a Japanese flag flying over Fort Canning was demoralising and troops milled about everywhere, most still fully armed. Some had almost no possessions other than the clothes they stood up in. The streets were choked with vehicles of all types and hundreds of private cars, abandoned by their owners, lined the roadsides. Fires still raged in bombed-out buildings, whilst the fire brigade stood helplessly by as the water supply had been cut by the invading Japanese.

General Yamashita ordered his troops to keep out of the city, fearful that they would run amok, but many chose to disobey his order and began to seize the spoils of victory. Tired and hungry, they entered houses and grabbed food, clothes and souvenirs, some paid for what they took but others simply took what they wanted. They severely punished the local population for looting, but in most cases turned a blind eye to the looting of their own troops. Drunk on victory, many of the Japanese soldiers raped any woman they could find.

In their book *This Singapore Our City of Dreadful Night* N.I. Low and H.N. Cheng wrote about conditions in Singapore in 1946:

*The raping varied in intensity. Some localities suffered more, some less. As was natural, it varied also according to the characters of individual soldiers. Some did it sadistically and brutally, booted and belted as they were, the Knights of Japan, without fear and without reproach; some did it indifferently as men answering mere calls of nature; some shame-facedly, mindful perhaps of the mothers who had borne them and the wives who were suckling their young in Japan. Their victims steeled themselves to accept the inevitable in as seemly a manner as their philosophy and good sense dictated. It would have been foolish to shout their disgrace from the house tops.*

*Chapter 4*

# Changi Prisoner of War Camp Singapore

On 16 February 1942, the Japanese ordered the conquered Allied commanders to move all their men and equipment to Changi, some 22 miles away in the north-east corner of the island. For the next three days there was a continual parade of men struggling along the highway towards Changi carrying anything they could find that might be useful for their future use. Frank and his Gordon Highlander colleagues set off on foot in sweltering heat, fully loaded carrying as much kit as they were able, to what many of them thought would be the end of their lives.

Trudging along, they passed scenes of utter destruction and chaos, wrecked houses and burnt-out buildings lined the streets, fly-blown dead bodies lay unburied everywhere. Water drained away down the monsoon drains contaminated with the blood, oozing the sap of the dead. Shops and cafés were boarded up. Black smoke from the oil installations drifted overhead. Singapore was completely devastated. Japanese soldiers sped around in stolen cars grinning at the defeated troops, some even demanded watches from the weary soldiers. Japanese flags now flew from many of the buildings as the local population tried to ingratiate themselves with their new masters. They were instructed to put their watches and clocks forward by one and a half hours to align with Tokyo time and the island was renamed Synon-Lo (the light of the south).

The people of Singapore lined the roads to watch the defeated troops marching along in silence. Mile after mile, the weary columns stretched along the roads towards Changi, each man immersed in his own thoughts about the future.

For the past ten weeks, my father had been trying to kill the Japanese invaders and stay alive himself, whilst Frank Pantridge had been trying to keep his wounded colleagues alive. Now they both had to come to terms with the fact that this 'live or die' situation was out of their hands; their fate lay with the conquerors. They had no way of knowing how long it would be before they would see their families again, or indeed could let them know if they were dead or alive. Rumours and counter rumours spread around the island. Some thought that they would be prisoners for quite some time, whilst others were convinced that with the Americans now in the war, it would only be a matter of months before the Japanese would be overrun and they would be freed.

Changi military prisoner of war camp should not be confused with Changi jail, where many of Singapore's expatriate civilian population were held by the Japanese. The military camp covered a large area of over six square miles at the eastern tip of the island, and it was to be the new home for around 53,000 men in barracks designed to house round 4,000. The sense of impotence and frustration by the men now in Changi was palatable.

As the Changi Peninsula had been the British Army's principal base area in Singapore for many years, the site had a well-constructed military infrastructure, including three major barracks – Selerang, Roberts and Kitchener – as well as many other smaller camps. The barracks had originally been built to house a brigade, usually around 4,000 strong, but it was now occupied by over 50,000 troops. A further large contingent was to be based at Seletar barracks a few miles away. The Gordons' old barracks at Selerang, now a grim blackened three-story building, had already been commandeered by the Australians, so they ended up in a hutted camp near the beach at Telok Paku.

For the first few weeks after the surrender, Changi was a haven of peace for the exhausted troops as there were no guards to worry about.

The Japanese were totally unprepared for the number of prisoners on their hands and the men were left to look after themselves.

My father said: 'In many ways, it didn't feel like a prison camp at all as we could move around freely within the perimeter. Our officers ran the camp in an organised military way and tried to make our lives as comfortable as possible during the early days of captivity.'

Army discipline was still very much in evidence though, as the officers tried to keep the men's morale and spirits up. It was decreed that officers should be saluted at all times and anyone breaking the rules was given extra duties or a day or two in solitary confinement – in effect being doubly locked up.

A weekly Regimental Sergeant Major's (RSM) parade was also set up, supposedly a morale booster, but this rule was universally hated by most of the men. Standards were to be kept high, boots polished and uniform smart, but of course this was impossible to impose as many of the men only had the clothes they were wearing at surrender.

Lieutenant General Percival advised all the prisoners in Singapore that it was their duty as soldiers to try and escape. He warned them though, that they must prepare thoroughly before making any escape attempt, to ensure that they had a reasonable chance. 'Your chances of escape are thin, you are on an island surrounded by enemy forces with the nearest friendly country many miles away,' he said. Most of them were aware that their chances of escape were poor anyway, and that recapture was likely to mean heavy punishment or death.

Accepting their imprisonment with stoicism, the Gordons set about making their camp as comfortable as possible. Boreholes were dug for water and sanitation and the men made the best of it during those early weeks of captivity. Anyone not knowing the truth might have mistaken Changi for a holiday camp in late February 1942, as the men frolicked in the sea from dawn to dusk. Initially plenty of food was available due to the large stocks held on the island, but water was more of a problem and had to be boiled before use to avoid cholera and dysentery.

Lieutenant Colonel Stitt did his best to keep the men of the Gordons together, but the Japanese had other ideas. They deliberately began to break up regiments and separate the men from their officers on the basis that by breaking up regiments, the prisoners were less likely to become an organised threat. Whilst quite content to let the officers run the camp with minimal interference, they made it quite clear that any man trying to escape would be shot. Such an edict was totally against the Geneva Convention of 1929, a convention which the Japanese had signed but not ratified. They had agreed to abide by its terms, but this promise was never carried out as many of the men were to find out.

With Singapore and Malaya now completely under Japanese control, escape from the island would prove to be very difficult. The Japanese navy ruled the surrounding seas, their air force was in control of the skies, with the nearest friendly peoples more than 1,500 miles away. Some men did however try to escape during the first few weeks of captivity with tragic results as we will find out later.

The victors were totally surprised at the number of allied prisoners they now had on their hands. On Singapore Island alone there were approximately 28,000 British, 18,000 Australian, 67,000 Indian and 14,000 local volunteers. Never in any previous conflict had so many British and Commonwealth soldiers surrendered to invading forces. It is not known exactly how many were taken prisoner, but it is now estimated that the number was likely to have been near 200,000.

A few weeks after their arrival in Changi, several platoons of the Gordons were ordered across the straits to Pengerang. They were to lift the anti-personnel mines that they had laid there some months previously. Such an order went totally against the Geneva Convention. Despite the protests of Lieutenant Colonel Stitt, they were ordered across the newly repaired causeway at gunpoint. This task was met by some apprehension by the men, as the mines, laid six months previously had become unstable. They were forced into the mine fields by armed guards to locate and lift the mines. The anti-personnel mine was a metal container shaped like a flowerpot, with 30 to 40 yards of trip wire attached. If the wire was tripped the pot would be ejected 2 or 3 feet into the air sending shrapnel flying in all directions. By now the mines and their wires were covered in vegetation and difficult to locate making the operation incredibly dangerous.

During the first day, Frank heard a loud explosion followed by a yell of agony. Rushing into the minefield, regardless of the risk, he picked up a young officer and dragged him out of the danger area. Despite his valiant attempts to save him, the officer named Stewart, died of his injuries a few weeks later back in Changi. Frank writes in his autobiography *An Unquiet Life*:

*Shortly after the mine-lifting operation started, there was an explosion and an agonised yell. I rushed into the minefield and picked up the young officer Stewart. He had horrendous injuries and died. In retrospect my action was crazy and suicidal. That I knew Stewart well was irrelevant. My action was reflex and unthinking. Stewart, the charismatic Battalion intelligence officer, was a most civilised character.*

Three months after the surrender, the situation in Changi began to change as Frank explained to a journalist after the war:

*Troops were cooped up in grossly overcrowded buildings. It didn't take long for dysentery to appear. Malaria rapidly rose because the drains had been disrupted during the bombardment. At first, we had a limited supply of drugs available, but these quickly became more difficult to obtain. Food also was now at a bare minimum, with only rice and a few bits of green vegetables. Under those conditions, it didn't take long for men to lose weight, and then the deficiency diseases appeared – beriberi, pellagra and scrotal dermatitis (called riceballs by the troops). A lack of vitamin B often caused painful feet. Many men's feet were particularly troublesome at night, resulting in them spending the night prowling around outside the huts. Others did not sleep at all because of the bed bugs. The evil, sickly smell in the hut sticks in my memory to this day.*

Wounded men who had been in the Singapore hospitals at the surrender were now transferred to the ex-military hospital in Changi, resulting in severe overcrowding. Many, who had been very badly wounded, died during the following months due to a lack of medical equipment and supplies. The Japanese left the military medical staff to look after the injured men with little or no resources. Frank did everything in his power to save the lives of the badly wounded and sick men, but he was fighting a losing battle. Had the Japanese provided even some of the basic equipment and drugs that were still available in the various hospitals in Singapore, many of those who died might well have survived.

Occasionally, some senior high-ranking Japanese officer would visit the camp and the men would be ordered to turn out in their best uniforms for inspection. Such visits were often photographed to demonstrate to the world how well they were looking after their prisoners' welfare. These were of course well-orchestrated publicity stunts. This infuriated Frank, as did the order that all ranks of the Imperial Japanese Army had to be saluted when they passed, if a man failed to comply, he usually got his face slapped. He invented his own style of salute by thumbing his nose, but they soon realised that this was an insult and he would get a crack on the head with a rifle butt.

By the end of August 1942, the Japanese became so concerned about the number of prisoners attempting to escape, that Lieutenant General Fukuei Shimpei, the new commander on the island, demanded that each man should sign a document undertaking not to try to escape. It read:

> *I the undersigned do hereby declare that on my honour that, under any circumstances, I will not attempt to escape.*

Almost everyone refused to sign the document and as a reprisal, on 2 September, all prisoners on Singapore island were ordered to march to Selerang barracks. Ironically, it was the barracks where the Gordons had been based when they first arrived in Singapore some five years earlier. Selerang barracks had been built to house 800 men, but it was soon crammed with over 17,000 men including the sick and wounded. The Japanese cut off the water supply, leaving only one tap to serve 17,000 men and there was no water at all for the toilets. Living conditions were dreadful and within a few days dysentery broke out, but still the men refused to sign.

Frank writes about conditions in Selerang in his book *An Unquiet Life*:

> *Each block had 1,800 men. Troops jammed like sardines on every floor, on the flat roof of every block and on the stairs and verandas. It was not possible to lie down. The Japs turned off the water supply to the blocks. Round the whole area was a barbed wire fence manned by sentries. Makeshift latrines were dug in the tarmac square, but these could not be reached by those incarcerated in the blocks. Dysentery was rife, and the place stank. Diphtheria and septic skin diseases were common.*

With the prisoners still refusing to sign the non-escape clause, Shimpei decided that an example should be set and four men who had been caught attempting to escape were sentenced to death.

General Homes, the most senior British officer protested vehemently and drew up a plea for leniency for the men that read:

*I have the honour to submit this earnest appeal for your consideration of the infliction of the supreme penalty of four prisoners of war who have been apprehended attempting to escape from this camp. I am aware that it has been made perfectly clear by you that any such attempt will incur the penalty of death and furthermore, there is no misapprehension on the part of the prisoners themselves that this is so. I will however take immediate steps again to impress on all ranks in this camp the inevitable outcome of any attempt to escape and would be that in the present instance you will exercise your clemency by the infliction of a less severe punishment than that of execution.*

*Signed E.B. Holmes, Colonel Commanding British and Australian troops, Changi.*

This plea was delivered by Captain Jones of the 122 Field regiment, Royal Artillery (RA) and his account of the delivery of the document is as follows:

*I arrived at the gaol soon after noon and the first person I saw was a Japanese interpreter. I explained my mission and stressed the fact that the plea for clemency was to be handed to the General. He wanted to take the papers – I refused to hand them over. He then brought a Japanese officer whose rank I could not make out as he had no insignia. I handed the paper to him. He appeared to speak some English. He read the plea for clemency, became very angry tore it up in pieces and threw them in my face. He then picked them up and stuffed them in my waist band. I then returned to camp and reported to Major Magee.*

At mid-day on 2 September, the four men were driven in a truck to Changi beach where four graves had been dug. They were made to stand with their backs to the graves and a firing party, made up of Sikh soldiers who had defected to the Japanese, lined up facing them. British and Australian officers pleaded for the men to be spared, but Shimpei was adamant they were to be shot. The men were offered blindfolds but all four refused. A Padre said prayers for the condemned men, each was given a cigarette and the officers saluted the men, who returned it bravely.

Lieutenant Okusaki, in charge of the execution, then ordered the squad to open fire. The four men were still alive after the first volley of shots, screaming out in agony, before they were finally finished off with a second volley of shots.

Bugler Arthur Lane of the Manchester Regiment was called upon to play the Last Post after the four men had toppled into their graves. I had the good fortune to visit Arthur at his home near me in Stockport, where he outlined to me his thoughts and feelings on that fateful day when the group were executed on Changi beach:

*When we arrived at the scene a large group of Japanese soldiers were already waiting for the 'carnival' to commence. A truck arrived from Changi prison. The back was dropped down and the prisoners were ordered to get down. One of the prisoners was wearing his pyjamas having been brought from the hospital. He could not walk or even stand on his own, so he was assisted by the Padre and one of the officers.*

*The prisoner's names and army details were read out parrot fashion. The three able men were taken to a position where five pieces of wood were protruding from the ground. Each one was placed in position in front of each pole and secured by rope. The fourth man (Waters) was carried to sit upright. There was a preamble in Japanese read out loud, which I believe was the sentence of death issued by the Japanese court, this followed with two Japanese officers performing some form of salute using swords.*

*There was a great deal of muttering coming from the English-speaking witnesses, with Colonel Wild stating that the men had not received a fair trial. Whilst this was going on ten soldiers were marched in line facing the prisoners. A Japanese officer shouted orders at which the ten soldiers lifted their rifles to the firing position. On the order to fire they opened up and all four men slumped forward to meet their maker.*

*Our own officers started shouting and swearing at the Japanese, who at gunpoint then ordered them to fall back. The Padre started to pray for the men, and I was ordered to sound the Last Post. I could not do it justice as I was so upset from what I had seen. I was trying not to show my tears and after a feeble attempt at the Last Post the Colonel ordered me to follow him.*

*After the war, Lieutenant Okusaki, the officer in charge of the execution was captured by a group of Australian soldiers who took him back to Changi beach and without trial, executed him in retribution.*

This was to be only one of the many Last Posts that Arthur Lane was to perform during his three and half years of captivity under the Japanese.

After the men had been executed, Lieutenant Okusaki turned to the Allied officers witnessing the execution and said, via his interpreter:

*You have witnessed four men put to death (two Australian and two British). They tried to escape against Japanese orders. It is impossible for anyone to escape as the great Nipponese are in all countries to the south and anyone escaping for here must be caught. They will be brought back here and put to death. You officers are responsible for the men under your command and you will again tell them not to go outside the wire. If they do, they will be put to death as you have just seen. We do not like to put them to death.*

Back at Selerang barracks the officers were by now only too well aware of the consequences if a serious epidemic broke out, the implications for the 17,000 men cooped up like battery hens would be catastrophic.

Another Japanese officer, Lieutenant Colonel Okamu addressed the officers shortly after the executions:

*I am Lieutenant Colonel Okamu. Up to now I have had no direct connexion with prisoners, as I have been in charge of arrangements to stop escape, for this reason I have not met any of you before. I am going to give you advice on the condition of prisoners. By looking at me you will see that I have had more experience in Army life by virtue of age. The reason for this meeting is that you have refused to sign the declaration forms. There must be strict discipline in the Army. I think you are to blame in not stopping trouble. The procedure is discipline and declaration. If you object, then you are to blame. You obey the regulations of other countries, then you must obey the regulations of my country. So, I think you will agree to our proposal and sign the declaration.*

*When you get your forms issued to you again there will be no cause for court martial. If you give praise to obey, you obey, and need have no fear of court martial. If you do not compel men to sign, they think they might escape. The four men were shot today, it was their own fault, but the responsibility lay on you their commanding officers. Looking at the miserable spectacle today, if you do not warn your men to sign then I fear officers have no capacity. If I were a CO in this group, I would take the whole responsibility and persuade them to sign. I would throw away property and reputation for my men. I believe officers must sacrifice for their men and men must sacrifice for their officers. I think officers must sacrifice something for their men in this event. All of you are mature in age. Do you still cling to property, I would say throw it away? I can't hope to persuade you to understand what I have said and these necessary things. How long will these miserable conditions last? At all events longer than you think.*

*Some of you expect troops will arrive for the USA and will come to rescue you. This will not be so, as the Japanese navy has superiority. The greater part of the United States navy is sunk and so is yours. Can you expect any troops to come and rescue you? I fear and hope that your hopes will not be fulfilled. How futile then any attempt to escape from here will be. You will find this my statement true when you return home and read about the Japanese navy in all the papers, propaganda wireless is untrue. Beginning in August 48 warships were sunk in the Solomon Islands battle and in the second battle more were damaged. From the Pacific to Africa we have command of the seas. So let me, a learned man, instil in your minds you cannot escape, and help will not come. It is a dream that will not come true.*

*So, you must consult and take responsibility to avoid miserable things happening to those who are innocent. Tonight, and tomorrow consult and do something to change yours and your juniors' minds as conditions are unhealthy. I am an old man and offer this good advice to you. I am now ready to answer any questions.*

Lieutenant Colonel Holmes acting as GOC replied making the following points:

*Our King forbids Officers to give parole and it is contrary to our nation honour to do so. Men had not been intimidated. It was made possible for everyone to sign. However, by 40,000 to 3 they were against it. The discipline of the Officers and men was excellent e.g. that march to Selerang was carried out by all units. Men had already been warned that in the present circumstances it would be foolish to try to escape. All PoWs would be willing to sign a form stating that they knew the penalty for attempted escape was death. Permission was asked to send a letter to the Jap General explaining the code of honour regarding this.*

Lieutenant Colonel Holmes thanked Lieutenant Colonel Okamu for having explained matters.

This appeal fell on deaf ears and Lieutenant Colonel Holmes had no option but to order the men to sign the non-escape document, he said to the men:

> *I am fully convinced that his Majesty's Government only expect prisoners of war not to give their parole when such parole is to be given voluntarily. This factor can in no circumstances be regarded as applicable to our present conditions, the responsibility for this decision rests with me, and me alone, and I fully accept it in ordering you to sign.*

On 5 September, the prisoners were released from Selerang and allowed to return to their camps around Singapore to their great relief. Frank recalled:

> *There was a rush for drinking water and a wash. The shower I had from a makeshift water pipe was one of the most pleasant experiences of my life. After fluid replacement I was delighted that I passed some urine. It felt like passing gravel. It may be that the kidney damage from the dehydration in those tropical conditions with 100 per cent humidity, was the cause of the hypertension from which I have suffered for the rest of my life.*

Lieutenant General Fukuei was tried by the war crimes commission at the end of the war, found guilty and sentenced to death by firing squad on Changi beach, the same spot as the four men he had killed more than three years previously. His sentence was later reduced to ten years imprisonment and justice was not done.

It was now more than six months since the fall of Singapore, and up to this time the prisoners had not been allowed to send any letters or messages home to their loved ones. Frank was desperate to let his family back in Northern Ireland know that he was still alive. With the Selerang incident settled, the Japanese relented and allowed them to send postcards home. The cards were very basic, with six basic messages to cross out, but at least sending them made the men feel a little better. What they were not to know however, was that many of the cards would take months to reach Britain, in some cases years. The Geneva Convention on the treatment of prisoners of war stated that:

> *Prisoners of war shall be allowed to send and receive letters and cards. If the Detaining Power deems it necessary to limit the number of letters and cards sent by each prisoner of war, the said number shall not be less than two letters and four cards monthly, exclusive of the capture cards provided for in Article 70.*

Of course, the Japanese completely ignored the convention once again.

As their captivity dragged on, the men in Changi had only one persistent thought in their minds, 'How long are we to be kept as PoWs and when, by some miracle, would we be freed.'

As 1942 came to an end, food became even scarcer. The mainly rice diet began to take its toll, with many men spending a good few 'unhappy hours' in the latrines with diarrhoea, constipation or dysentery. Vitamin deficiency was also becoming a major problem. For Frank, rice meant a sweet rice pudding with a baked skin on top, served up in the RVH canteen back in Belfast and he found it difficult to adapt to the diet.

Gardens were started all around the camps, but with a shortage of water and so many mouths to feed, the meagre vegetables that did survive did not go far. My father said about his time in Changi: *'We were living on a rice diet and very little else, everyone was beginning to feel the lack of vitamins and I was losing weight rapidly.'*

Dysentery was now reaching epidemic levels and proper sanitation was crucial if it was to be contained. Borehole drilling equipment, used by the postal service for erecting telephone poles, was available and the fittest men in groups of five or six would drill holes to a depth of at least twelve feet to serve as latrines. A wooden box with a hole in the middle as a seat was erected just above ground level. Poor washing facilities and little soap made things worse.

Soon many of the men in Changi began to die. During the first seven months of captivity more than 400 died, some from their wounds during the fighting and others later, from diseases like dysentery, beriberi and diphtheria. Frank and his fellow medical officers did their utmost to keep the men alive, but with a scarcity of medicines they were fighting a losing battle.

To try and combat the lack of vitamins in their diets, they came up with the idea of making a vitamin B drink where ground rice, sugar, salt, ground peanuts and yeast were mixed together in large beer bottles and left to ferment. Such a concoction saved many lives. They also invented an apparatus that extracted black juice from Lalang grass, but the men found the result evil tasting, it made them retch and despite its health-giving properties, many refused to drink it. One of the natural resources available around the camp was the hibiscus bulb. Its bell-shaped flame-red flowers grew in abundance in the hedgerows and, when its leaves were boiled, it made a very appetising and nutritious soup. They were also able to forage a few papaya and coconut palms along with some Malay cherries to add to meals.

Every prisoner was entitled to receive Red Cross parcels sent out by the Red Cross in Geneva, but the Japanese kept many of them for themselves, with the prisoners getting very few. It was not until September that they were given a few tins of food, a pair of boots each and a hat.

The Japanese now began to set up working parties in an attempt to get a devastated Singapore back to some sort of normality. The most popular parties were those sent down to the docks, where there was ample opportunity for the prisoners to steal food. As Frank had come into conflict with senior officers in Changi and they wanted to get him out of their hair, he was sent on one such party. Despite the filthy camps near to the docks, the men were able to steal a variety of foodstuffs stored in the 'godowns' (warehouses). The godowns were stacked with food and drink but of course stealing was not without its risks, as they quickly realised. Their captors strictly enforced a regime of corporal punishment if caught stealing.

One of the camps near the docks had an open sewer running right through it and one man had the misfortune to fall into it whilst going to the benjo during the night. According to Frank '*the place swarmed with faecal-feeding flies transmitting dysentery*', a phrase that he had remembered from a lecture he had attended back at QUB given by Sir William McArthur.

Some men on these early working parties into Singapore town had the misfortune to witness some of the worst atrocities carried out by the Japanese. They observed first hand, some of the dreadful massacres of the Chinese population by soldiers who seemed to have only had only one thing on their minds, to exterminate them. After being rounded up and interrogated by the Kempeitai (Japanese military police), they were obliged to hand over all their personal possessions, before being taken in captured British lorries to Tanjong Pagar Wharf or Changi beach, where they were beheaded, bayoneted or machine gunned. Others were roped together and taken on barges out to sea, where they were thrown overboard.

One of the few advantages of Changi for the prisoners was that it is right next to the sea. They could go swimming every day to get some relief from the heat and to wash their sweaty bodies. Although not a great swimmer himself, Frank did take advantage of the opportunity to relax in the salty water. It was a hammer blow when, in late 1942, the Japanese put the beaches out of bounds. No reason was given for such an order, but it was assumed that they wanted to apply as many hardships as possible to men who had surrendered. Despite vigorous protests from the officers, the order was upheld and anyone caught swimming was severely punished. Not wanting to give up one of the few pleasures available to them, quite a few men crept down to the beach after dark for a quick dip in the warm waters of the Singapore Straits.

To alleviate the boredom, other activities were quickly arranged including a variety of sports and entertainment. Changi already had an open-air cinema which was soon turned into a theatre, where a group of men with experience of the stage put on many excellent productions. Religious services held by the various Padres were always well attended as many men held strong religious beliefs. Holy Communion was regularly conducted using fermented wine made from berries collected from trees, or 'Chateau Changi' as it was labelled by the men. Changi village itself had a well-stocked village library where a ready supply of books was available. The school and garage were also used for lectures and discussions.

Rumours had been circulating around Changi from early March 1942, that the Japanese were planning to send prisoners up into Thailand to help build a railway. Whilst radios had been confiscated after the surrender, several clandestine sets (known as canaries) had been hidden away and were now secretly operating. Any snippets of news received were quickly circulated and the men soon learnt that the Japanese had taken Rangoon, the capital of Burma, and were pushing towards the British colony of India. With such a swift advance up through Burma, their supply lines had become stretched. To reach the front-line troops in northern Burma their ships had to make the dangerous journey from Singapore up to Rangoon via the Straits of Malacca. By this time, the American Navy was very active in the Andaman sea, with their submarines posing a huge threat to Japanese shipping steaming towards Rangoon.

The only safe and logical route for supplying their front line was overland from Bangkok to Rangoon. The road network between the two cities was poor and the only existing rail link was a 40-mile stretch of track running from Bangkok to a small town in Thailand called Ban Pong. The Japanese decided that their only option was to build a railway that would link Ban Pong with the Burmese town of Thanbyuzayat some 250 miles north-west, from where a rail link already existed on to Rangoon. It was rumoured that such a rail link had been considered by the British many years earlier, but it had been deemed just about impossible due to the difficult terrain. Mountains up to 5,000 feet high and miles of dense tropical jungle made such a project virtually impossible and the scheme was scrapped. However, the foremost authority on the Thai/Burma Railway, Rod Beattie is sceptical about such claims that a scheme had been even considered by the British: 'Years of research by a number of people have failed to locate any evidence of British plans for a railway along the River Kwai and through the Three Pagoda Pass into Burma,' he asserts.

With all of Malaya now in their hands and with the Thai government under their control, the Japanese pushed ahead with this project despite the huge logistical difficulties it posed. After all, they had 137,000 'willing' and 'disposable' labourers on their hands.

The rumours circulating around Changi soon proved to be true and work on the railway started in May 1942. The first group of 3,000 Australian prisoners left Singapore on 14 May in the *Celebes Maru* and *Toyohashi Maru* bound for Moulmein, Burma. The ships picked up 450 British prisoners, led by Captain Dudley Apthorp of the Royal Norfolks, in eastern Sumatra before arriving into Moulmein in Burma on 24 May after nine days at sea. This first shipment of prisoners was designated as A Force.

When it was first announced to the men in Changi that many of them were to be transported up into Thailand to work on the railway, quite a few were excited about the prospect. The Japanese, through their interpreters, painted a rosy picture of excellent living and working conditions, plentiful supplies of food, excellent medical facilities and that they would be well cared for. They were told:

> *The climate of the new location would be similar to that of Singapore, the men would be distributed around seven camps, all would be in pleasant and healthy surroundings, sufficient medical personnel to staff a 300 bed hospital could be sent, as many blankets and mosquito nets as possible were to be taken, a band could accompany each group of 1,000 men, canteens would be established within three weeks of arrival, no restriction would be placed on the amount of personal gear taken, tools and cooking gear were to be taken and transport would be available to take heavy goods and there would be no long marches.*

Most of the prisoners were happy to get away from the boring crowded Changi camp and looked forward to moving up into Thailand.

The following month, the next group of 3,000 prisoners left Changi for the railway, this time by train. Their job was to prepare the way for the arrival of the main force due to leave in October when the monsoon season had ended. My father left Changi on 25 October 1942 for the railway, and by the end of September, around 18,000 prisoners of war had been transported north by the Japanese.

Back in Changi, Frank watched many of his colleagues leaving on a daily basis, but as 1942 came to a close, he was not detailed to go north on any of the parties. Although the terms of the Geneva Convention 1929, stated that prisoners of war were not to be used on dangerous tasks or those helping the captor's war efforts, the Japanese rode roughshod over it. Their premise was that any soldier surrendering had forfeited all rights, so were set to work on any tasks that helped their war effort.

By March 1943 and with the war now turning against them, the Japanese decided that the completion of the railway was a priority.

It was now only twenty per cent completed, and the original finish date of December 1943 was brought forward to August. To meet this target, a 'speedo' period was implemented. The prisoners were pushed even harder, often working twelve hour shifts day and night for seven days a week. This was now going to be a mammoth undertaking and they began to transport 19,000 more PoWs and tens of thousands of Asian labourers to try and complete the job.

When these men began to arrive in Thailand, the monsoon season was well under way. The Japanese had intended to suspend or reduce work on the railway during the monsoon season, but with the brought forward completion date, this was now disregarded much to the discomfort of the prisoners. It was at this point in time that the 'tragedy' of the railway began to unfold and Frank Pantridge's new 'hell on earth' started.

*Chapter 5*

# F Force and the Death Railway

On 28 April 1943, Frank's time in Singapore came to an end when he joined 7,000 men preparing themselves to be transported up into Thailand. His party, to be known as F Force, was made up of 3,600 Australians and 3,400 British prisoners. One in every three men in the group were unfit to march or work, but the Japanese were so intent on providing labour to finish the railway, that they simply ignored requests from the senior allied officers to leave the very ill behind. Although they did not know it at the time, they were bound for the infamous Nikki/Songkurai area, the most northerly of the series of camps in Thailand.

The men were given inoculations, but of course they had no idea what they were being injected with and indeed how effective the treatment was likely to be. Despite constantly asking the Japanese what they contained, the medical officers were not told anything. As a doctor Frank asked what was in the vaccine but they simply ignored his requests. Padre Duckworth, who was to be one of the five padres on the force, said:

*The Japanese told us we were going to a health resort. We were delighted. They told us to take pianos and gramophone records. They would supply the gramophones. We were overjoyed, and we took them. Dwindling rations and a heavy toll of sickness were beginning to play on our fraying nerves and emaciated bodies. It seemed like a bolt from the tedium of life behind barbed wire in Changi. They said, send the sick, it will do them good. We believed them, so we took them all.*

Major Hunt of the AIF was very sceptical about the Japanese promises and spoke to the men prior to leaving:

> *We have been told that we are going to a convalescent camp somewhere up north, the commanding officer believes it, but I don't. We will be going to a land where disease will be rife so prepare yourselves for the worst – you will encounter diseases you have never heard of and I fear for the future.*

How prophetic his words turned out to be!

Many of them took too much kit with them; they left burdened with large packs, bed rolls and even suitcases, a group of bandsmen even tried to take a piano with them. F Force was supposed to include a medical party of about 350, of which Frank was one. They packed enough equipment to set up a central hospital for about 400 patients, with a medical supply that would last for three months. As it turned out, the force of 7,000 men had a mere thirty-four medical officers to look after them with very little medicine or equipment. Lieutenant Colonel S. W. Harris RA, a celebrated sportsman having represented South Africa at tennis, England at rugby and boxed in the Olympic games, was to be the commander. Thirteen trains were to transport the 7,000 men from Singapore to Ban Pong in Thailand.

The train sitting in Singapore station that would transport Frank and his colleagues 1,100 miles up country was made up of a series of metal cattle trucks around 20 feet long. Each one was 7 feet wide and 8 feet in height. There was only one entrance/exit to each truck via a central sliding door about 6 feet wide. Thirty men, along with their kit were crammed into each truck, leaving each man only about 5 square feet for himself.

It took some time for Frank to get himself organised. The only way he could make himself comfortable was by sitting on top of his kit, lying down was impossible. Stops were only made when the train needed fuel and water. As the stops were quite short, they had to make a quick dash for the water towers along the line to fill their water bottles and to try and wash away the dirt from their bodies. They had to relieve themselves along the side of the track, often in full view of the local people. The only food they had to eat during the journey were the cooked rice balls given to them at Singapore station.

Frank recalled the journey:

*The only latrine facility was the use of the sliding doors on one side which had a chain stretched across the opening. The PoWs had to hang precariously on the chain as the train lurched along. Those with acute dysentery were passing up to twenty motions a day. Some of the men were too weak to hang out of the door unaided and had to be held in position by their comrades. The dysentery cases weren't allowed off often enough, there was excrement everywhere and men simply wallowed in their own and other people's shit and piss for four days and nights. It was baking hot during the day and bitterly cold by night and dysentery had by now got a grip on many of the men.*

After crossing the border from Malaya into Thailand, the train eventually arrived into Ban Pong station after a journey of 1,100 miles taking over forty hours, much to the relief of the exhausted and hungry prisoners. The Japanese guards who had accompanied the party on the train, handed them over to another set of guards waiting at Ban Pong. They were ordered to line up for inspection on a piece of waste ground adjacent to the station and to lay out their kit for inspection. The guards then proceeded to simply help themselves to anything they fancied, before marching them to Ban Pong camp.

Ban Pong camp was aptly named as it was a hell hole, the smell was appalling. The monsoon season was well under way when Frank and his colleagues arrived, with the rain falling in torrents and clouds hanging so low that the men felt they could reach up and touch them. The whole camp was a huge morass of mud, with bugs and insects of all shapes and sizes crawling over their bed spaces. The huts were built of bamboo, with traditional attap (thatched) roofs and floors that were just a sea of mud and maggots. Cockroaches, ants and spiders crawled out from the cess pits that were used as latrines. Large bluebottles hummed and buzzed everywhere – a fertile breeding ground for diseases of all kinds. Their gear, including vital medical supplies was piled unguarded in a heap near the station, resulting in most of it being either lost or stolen. The grand piano was sold to a local café so that money could be obtained to buy food from the local traders.

Their final destination was to be Nikki–Songkurai about 180 miles further north. Whilst they were promised transport, this was not made available and they were to be making the journey on foot. These men, who had been captives for over a year, existing in Changi on a diet that was quite insufficient and unsuitable for European stomachs, were now expected to march 180 miles. They were told that they would be marching overnight to avoid the searing heat of the day.

Frank was glad to get out of the crowded hell hole that was Ban Pong and he lined up clutching his medical bag and carrying his paltry belongings just as the sun was setting. The first two stages of the march north were on a loose metal surface, but this soon fizzled out into a narrow elephant track through dense mountainous jungle. The men fought their way along up to the ankles in thick glutinous mud, marching for around ten hours each night, with only a stop of around two hours for water and a meagre meal at the half-way point. Any man that dropped behind the group was likely to be set about by local Thais and robbed. Around twenty men disappeared during the march, possibly murdered by Thai bandits or simply dropping by the wayside and dying of exhaustion.

For some reason the Japanese took torches off men who had them, making it even more difficult to see and find their footing. One officer wrote:

*At times in the darkness they could only see the man immediately in front by his enamel plate or mug fixed to his pack. Sometimes white pieces of cloth or towels were hung over people's shoulders so that we could see them and so keep close together. Often, they slipped in the ruts and went sprawling in the mud and trekked for miles over the ankles in wet mud.*

It was very difficult for the men to get any sleep during the day as often the Japanese guards insisted on the men joining working parties already there. There was little in way of huts or shade in the staging camps and they would simply curl up under a bush or a tree to try and get some sleep. One man who was already in a camp on the route recalls seeing the men of F Force arrive:

*Sometimes we would see these men lying in the scrub near our camp. They were poor-looking creatures and there wasn't a thing we could do to help them.*

The medical supplies they had brought with them from Singapore, lagged miles behind the parties and often never caught up with the groups at all. Ulcerated and blistered feet occurred in large numbers of the men which the medical officers tried to treat as best they could at the various stops, using only banana leaves and bandages from sleeves and legs of pants. Frank and his fellow medics were selfless in their efforts to look after their men, often doubling back to help stragglers and encourage them. He tried his best to warn the men of the dangers of drinking water straight from the rivers and streams. Most of them were so thirsty that they ignored his pleas, with many ending up with dysentery. The men would say, 'it's OK, sir, it's running water', demonstrating their faith in the sterility of running water despite Frank's pleas. With the monsoon season in full swing, it was difficult to light fires and to keep them alight, in order to boil the water.

On 9 May one of the Aussies shouted out 'It's Mother's Day, lads.' This of course raised the morale of the party as their thoughts drifted back to their loved one back in Britain and Australia. Many decided to pick some white jungle flowers and tie them onto their packs – a Mother's Day that was never forgotten by those who survived the ordeal.

As the days and nights slipped past, the group got weaker. Some began to throw things away to lighten the load, blankets, towels and items of clothing were cast off, leaving them with just shorts, shirt, a mess tin and a water bottle. The Japanese guards showed scant mercy to any men who fell by the wayside, prodding them with bayonets and threatening to shoot them if they did not get to their feet and march on. The weaker men were helped along by their more able colleagues.

F Force marched for seventeen nights, covering an average of about 13 miles each night, stopping during the day at transit camps, most simply roadside clearings in the jungle with very little cover from the sun. Food was simply a handful of rice or onion stew with warm water to drink, a totally inadequate diet for men who had been undernourished from the start. Although F Force passed through camps where prisoners who had been sent up to the railway earlier were living, they were prevented from making any contact with their colleagues, supposedly to prevent the spread of disease to the new arrivals.

Camps had been built along the proposed route of the railway at regular intervals, with the distance between them varying depending on the difficulty of the work. On the flat plains the camps were about four miles apart, whilst in the mountains they were much closer together. In many cases they were built along the riverbank to allow supplies to be delivered by barges, but in these camps, for some reason the Japanese prohibited bathing in the river, an order that was largely ignored by the hot and sweaty prisoners.

On 21 May, after seventeen nights of marching since they left Ban Pong, F Force eventually arrived into their destination, the Nikki-Songkurai area near the border with Burma. Around 1,000 men had been left behind at staging camps or had died en route. They were now effectively cut off from the rest of the world.

There were five camps in the area, Nikki the HQ and hospital camp, Shimo Songkurai, Songkurai, Kami Songhurai and Changaraya. On arrival, they found their accommodation was in unfinished huts. Cholera had broken out amongst the men, but Frank's medical team lead by Australian Major Bruce Hunt did their best to combat the disease. Frank himself was now in a bad state, having suffered several bouts of malaria, but he became obsessed with trying to limit the spread of cholera, setting up saline drips using distilled water. With scant regard for himself, he did the best he could for the men in his charge with no medical supplies or drugs. It was simply a matter of trying to keep their morale up and ensuring that they had the best possible hygiene. Diphtheria also proved to be a major concern for the medics, as they had no means whatsoever of treating it other than keeping the patient flat on his back for several weeks and not moving. This kept the strain on the heart to a minimum, an organ that Frank was particularly interested in. They were forced to just let it run its course.

Bed pans were carved out of bamboo and needles for intravenous injections were made from bamboo tips. Stretchers were constructed out of bamboo as were buckets; clothing was cut up to make bandages and men were cajoled into taking in a large fluid intake. Intravenous saline, a mixture of boiled river water and table salt, was heated to blood temperature and injected.

At Nikki-Songkurai they were set to work on clearing jungle, cutting down trees, driving piles, carting earth and moving rocks.

F Force was now spread over the five camps along a 25-mile stretch of the railway; camps that were to turn out to be some of the worst on the entire railway. Japanese engineers were in charge of the workforce now and they proved to be what Frank called 'a most evil breed'. They were ruthless in their pursuit of the completion of the railway regardless of the cost in human lives.

In Frank's camp there were around 175 men who were identified as cholera carriers and they were quickly moved to an isolation camp. It was a pitiful place, located in a sea of mud where they had nothing to sleep on but their own kit. The medical officers were often beaten by the guards for trying to prevent very sick men from being ordered out on working parties. Reveille was sounded every morning at 0600hrs when it was still dark, rousing the men from their troubled sleep. Often, they would awake to the sound of rain hammering down on the roofs of their huts. Breakfast was a watery mess of rice with a few beans thrown in, along with a mug of weak tea. Tenko (roll call) was at 0700hrs, when the men had to stand in the rain outside their huts to be counted before marching to their place of work on the line.

When the monsoon set in with a vengeance on 22 May, the driving rain came down incessantly for over two weeks without a pause. It seeped through the atap roofs of the huts, soaking the men as they tried, usually in vain, to sleep on their wooden cots. During this period the river rose in flood, completely swamping several of the camps. Every unmade road and jungle track turned into a sea of glutinous mud, and water lay in great stagnant pools. Many men died of pneumonia during this dreadful period at the end of May, as they were never able to get properly dry, returning to camp each day soaked to the skin. Awaiting them was a meagre ration of watery rice that turned even more watery if not eaten quickly.

Clifford Kinsvig says in his book *Death Railway*:

> *The monsoon broke proper on 22 May and for the next 16 days the driving rain lashed the railway trace almost without pause. The Kwai Noi rose in flood and swamped several of the jungle camps completely. Every unmade road and jungle track became a watercourse and a layer of thick glutinous mud lay on the surface of even the higher camps. The rain found holes in even the best constructed atap huts and it poured through the desperately inadequate tents which were all that the prisoners had for housing. They were never dry: after a rain-soaked day's work they would return to a meal of watery rice which the rain diluted even further. Then retire exhausted to a wet bamboo bed or a sodden canvas one.*

As well as cholera, malaria became a major problem. Although usually not as fatal as cholera, it proved to be a real headache for the medical officers. The Japanese did not provide any mosquito nets and quinine was rarely available. In Nikki/Songkurai camp, out of 1,600 men, around 1,200 died in eight months from May to December 1943, with only 182 still alive when the railway was completed. Cholera was the main cause of death and Frank overheard a couple of Australians discussing it, '*What's this cholera mate?*' '*You shit, you vomit, then you're dead*,' his friend replied.

Initial signs of cholera were 'rice water' stool, livid cyanosis (a bluish discolouration of the skin), vomiting and pains in the gut. Finally, the body temperature drops dramatically and death soon follows.

One soldier said: '*I saw one medical orderly wipe a cholera victim down, forget to wash his hands and smoke a cigarette – he was dead within forty-eight hours.*'

Even though some of the men were inoculated with vaccine from the limited stocks they had brought from Changi, cholera spread through the camps like wildfire. Someone in the bed next to you would be alive when you went out on a 'Speedo' work detail in the morning and by the end of the shift, his bed space would be empty. The only treatment they had available was to administer large quantities of saline fluids using nothing more than primitive equipment.

When F Force arrived, the huts in the camps had not yet been roofed, the prisoners themselves setting to work covering them. Frank describes the living conditions in the camp at that time:

*We were housed in long bamboo huts covered with a thatch made from nipa palm. Inside, there were bays about 14 feet wide, where four to ten men were to sleep. Each day before dawn, the men were awakened, bolted down boiled rice and then split into gangs. Many were by now barefoot and clad only in loincloths which we called Jap-nappies.*

With his medical background kicking in, Frank escaped cholera by fastidiously boiling his water and indeed everything he ate was heated to as high a temperature as possible on the open fires. Their rice came 'polished' with the husks (the rice polishings) that contain vitamin B removed. The Japanese refused to give the large quantity of husks to the prisoners as they deemed them worthless, despite the potential for saving lives.

With a lack of vitamin B in their diet, many men succumbed to cardiac beriberi. There are two main types of beriberi: wet beriberi and dry beriberi. Wet beriberi affects the cardiovascular system resulting in a fast heart rate, shortness of breath and swollen legs. Dry beriberi affects the nervous system resulting in numbness of the hands and feet, confusion, trouble moving the legs and pain. In wet beriberi, the victim becomes bloated and his heart becomes enlarged, resulting in possible heart attacks. Vitamin B1 deficiency causing diseases such as beriberi were already widespread in the Far East before the Second World War and medical officers like Frank were very aware of this.

The Japanese had set out a ration scale for the prisoners, but in reality, what they actually received was nowhere near that amount:

Figures in grams per day: Rice – 660, Vegetables – 500, Meat – 150, Salt – 5, Tea – 5, Condiment – 5, Sugar – 20

At Nikki/Songkurai food and medical supplies were brought in by lorries from Burma via the Three Pagoda Pass. During the monsoon season however, quite often the road was impassable, and the men had to walk over 12 miles to collect supplies by hand and carry them back to camp. By the time they got the small amounts of meat or fish they collected back to camp, it was often rotten and full of insects due to the hot and humid environment. They were so hungry that they would cook the food anyway and skim off the insects before serving it.

The Japanese thought that many men were feigning sickness to avoid working on the railway, and to force them out to work, they adopted a policy of 'no work no food'. A policy that resulted in their health deteriorating even more.

Another evil the men had to contend with was 'rice balls' – not a Chinese delicacy, but the skin of the scrotum becoming raw and red. It eventually split and peeled off completely due to the lack of Vitamin B2. 'Happy Feet' was another symptom of the lack of vitamins, when a searing pain would shoot through the soles of the feet, often in the middle of the night. Many men would be up walking about in the huts for hours on end to try and get some respite.

Most camps on the railway had their own 'hospital' hut staffed by medics like Frank, but with little medical supplies, they had little to offer the men other than support and encouragement. Frank tried his best to prevent the really sick from being driven to work and along with colleagues he was beaten by the Japanese guards. The absence of protein was a major problem and they tried ingenious methods to obtain it from jungle sources.

Many operations were carried out under extremely difficult conditions, with efforts ranging from a simple appendectomy to the major amputation of a leg. Trench foot, a medical condition caused by prolonged exposure of the feet to damp and unsanitary conditions was also a big problem for the medics. Jungle ulcers were another major problem, as the smallest scratch on the skin would steadily develop into an ulcer that would eat away at the tissue until the bone was exposed. With no drugs available, their only option was to scrape out the rotten pus with a sharpened spoon, a procedure that was incredibly painful for the patient who often had to be held down and a damp rag stuffed into his mouth. Amputations were often carried out using no more than an ordinary meat saw, with patients often dying of shock, due to the lack of any anaesthetic. If they were lucky enough to survive the operation, they more often than not died of infections to the wound. Successes were known however, and men wearing artificial limbs made from bamboo, became a common sight.

Another crude method used to combat jungle ulcers in the camps near the river was to hang the infected leg in the river, where fish would nibble at the rotten flesh helping to cleanse the wound. Frank and his colleagues had to work with whatever methods they could to try and save lives.

Many of the men who had a variety of peacetime skills and trades put them to good use. They improvised stills to make saline solution from old petrol drums with tubing stolen from Japanese vehicles, retractors made from spoons and forks for the operating theatres. Blood transfusions were done through a doctor's stethoscope with blood donated by the healthiest of the prisoners. Frank Pantridge and his fellow medical officers performed miracles with the scant resources they had available and undoubtedly many lives were saved by this dedicated group of men. Despite their efforts though, between May and August 1943, over 3,000 men died from dysentery, malaria and cholera.

After only a few months on the railway Frank himself developed cardiac beriberi, his heart enlarging to three times its normal size. He was unable to even undertake basic medical supervision and spent most of his time lying on his bed.

In Takanun camp, some 45 miles south of Songkurai, Lieutenant Colonel Pond of the Australian Infantry Brigade was so appalled at the condition of the men that he wrote this verse in his diary:

*When your stomach turns to water and your feet begin to rot*
*When your outer skin is frozen but your blood is bloody hot*
*When your limbs begin to swell*
*And your ulcers burn like hell*
*Then you can say, but only then*
*You've joined the F Force, Thailand men.*
*When malaria has hit you or the Thai typhu comes to town*
*When dread cholera stalks behind you and the rod test bends you down*
*When the sandflies bite your ears*
*Or a 'basho' Nip appears*
*Then you can say, but only then*
*You've joined the F force, Thailand men.*

At the beginning of August 1943, Lieutenant Colonel Francis Dillon took charge of F Force. When Dillon arrived, he was appalled at the state of the camps where 800 men were in hospital at Songkurai. Dillon was to become a hero to the men in F Force by boosting morale and enforcing strict discipline in the camps.

By now the Japanese had decided to move the worst cases to a camp called Thanbaya in Burma, where they said that food would be easier to supply. The cynical among the prisoners reckoned that it was just an excuse to get rid of the terminally ill to a death camp and that the journey itself might be a death sentence.

As Frank was in a bad way and still suffering from beriberi, he was transferred in the back of a truck to Thanbaya at the end of August. Despite his serious illness, he was determined that he would not die, even though his companion in the next bed did his best to drag him down into the depths of despair. Each morning this man would call across to him: 'another bloody hopeless dawn, Frank.' The word hopeless was a feature of the man's vocabulary throughout the day and Frank did his best to ignore him, whilst repeating to himself: 'I will not leave my bloody bones in Burma.'

The death rate in Thanbaya was high and men who passed away during the night were simply slung onto bonfires the next morning and cremated. Frank recalled: *'As the bodies on the bonfire incinerated, the skulls made explosive noises. Every morning I could hear the repeated banging of exploding skulls.'*

The medical staff of F Force were fighting a losing battle against diseases in the camps all up and down the railway. In Thanbaya camp, the medical records show that by September 1943 out of approximately 1,800 men, around a third of them had beriberi, a third had dysentery and another third had malaria. Most of the rest had tropical ulcers.

Major Bruce Hunt of the Australian Medical Corps was so concerned about the health of the prisoners, that he wrote to the Japanese medical officer in charge of Thanbaya camp:

*The health of the camp as a whole is getting worse, not better, every day. The increase in beriberi being the chief cause under present circumstances and in the absence of immediate supply of very large quantities of vitamin containing foods, drugs and dressings I anticipate not less than 500 more deaths in the next six weeks. To these must be added over 280 deaths, which have occurred in the past six weeks. In twenty years of medical practice and after extensive experience of two wars, I have never seen men in a more pitiable condition of health than the men in this camp. The real*

> *tragedy lies in the fact that much of the disease is really curable if proper vitamin-containing foodstuffs, such as beans and towgay and proper drugs were made available in sufficient quantity and I earnestly impress upon whatever authority this may reach that, through no fault of their own, men are dying in hundreds and will continue to die until help comes.*

Despite their personal dreadful conditions, many of the prisoners spent time worrying about their loved ones back in their home countries. War with Germany was still raging and they were able to find out news through the illicit radios that were operating in some of the camps. The radios, called canaries by the men, had been built using materials stolen from the Japanese and were hidden in many ingenious places. The penalty for being in possession of a radio was death, so their location was kept secret. Very few prisoners knew who had them and news obtained was passed around by word of mouth. In one camp two British officers were beaten to death with pick handles for being caught in possession of a radio.

The Japanese had issued cards to be sent home, usually preprinted so that they could cross out what was not relevant *'my health is (good, usual, poor), I am ill in hospital, I am working for pay, I am not working, best regards.'* A small space was left at the end for a personal message. Most of the cards took over a year to reach home via the Red Cross in Switzerland. In many cases men had died before their cards reached their loved ones. Mail from their families and friends back home rarely reached the camps, particularly on the upper reaches of the railway. In some cases, this was used to blackmail the prisoners to make them work even harder. By September 1943 some 470,000 letters were waiting to be distributed to the camps.

During their captivity the men did their very best to sabotage the Japanese efforts to build the railway. They understood that anything they could do to slow down the Japanese advance into Burma, would be beneficial to the Allies. On his return home my father wrote in his release questionnaire '*we often left nuts loose on equipment so that wheels would drop off.*'

Whilst conditions were dreadful in the many camps along the railway, there were always some lighter moments. In one camp, the Japanese commander was so concerned about men dying, that he decided to get the prisoners to write wills so that their families would know at the end of the war. This of course amused the men as most of them had very few belongings to leave. One man wrote in his will: '*I leave to my mother and father Woolworths, Marks and Spencer, Freeman Hardy and Willis and the Ritz hotel.*' Of course, this caused great amusement in the hut when he told his comrades that night. Many years later when this man was making a proper will back in England, he was asked by the solicitor if he had ever made a will before? '*Indeed, I have,*' he said, '*in a prisoner of war camp on the Thai/Burma railway.*' The solicitor found this hilarious and asked him if he still had it. He then added an amendment to the bottom of the new will to the effect, '*I hereby revoke all former wills and dispositions heretofore made by me.*'

Amazingly, but at a terrible price, the Thai/Burma railway was finally completed ahead of schedule on 17 October 1943, when track-laying parties from both the Thailand end and Burmese end met at Konkoita. Over 150 million cubic feet of earth had been moved and nine miles of bridges built during the previous sixteen months. The prisoners in Konkoita were lined up to watch a Japanese general hammer a golden spike (at least that is what the men were told at the time, but the spike was actually made of copper not gold. The actual spike is now in the Death Railway museum in Kanchanaburi) into the track and a C56 locomotive was driven ceremoniously along the section of line.

In honour of their efforts they were each given two ounces of soap, ten Thai cigarettes and fruit, items that had probably come from Red Cross parcels. Wreaths were laid on the line by the Japanese in honour of the allied prisoners who died on the railway and an officer made a speech to the men:

> *Thank you very much for realising the importance of the railway to Nippon. We are sorry some of your friends have died but it is the duty of the soldier to die when necessary. Soon you will be going back to Kanchanaburi and Non Pladuk. Continue to be good soldiers. Take care of your health so that when peace comes you may return to your own country, to your dear wives and children and to all your dear friends.*

To celebrate the occasion, the Japanese decided that a football match should be arranged between the prisoners and the guards. For the emaciated men, this proved to be a difficult task, but as football was Britain's national sport, the men were delighted to take on their captors. The result of the match is not recorded.

A few days later, the first train load of Japanese troops bound for the Burma front passed through Konkoita, much to the delight of the Japanese engineers. During the following weeks, twelve trains a day loaded with troops and supplies creaked and groaned their way north into Burma where the Japanese were preparing to attack the British forces on the Indian border. The train crews would not have been aware of the cost in human lives, as they chugged along over the skeletons of the dead along the line. They would be oblivious to the ghosts of the prisoners who had died; the Japanese had achieved their objective in building a railway with a display of callous cruelty and complete disregard for human life.

Up in Thanbaya, Frank was still clinging onto life, with no knowledge about the completion of the railway. With the railway finished, the Japanese decided that the very sick prisoners should now be moved down to Kanchanaburi by train. Frank was carried from his cot by the medical orderlies down to Thanbaya station where he was carefully loaded onto a train where at least the carriage had wooden benches. He was glad to be leaving the dismal surroundings of a camp that had seen so many prisoners die whilst he was there. His journey south was slow, and often disrupted by frequent derailments due to sabotage carried out by the prisoners during construction. When his train reached the Wampo viaduct that hugged the cliff face above the river, those men who were able, looked down at the sheer drop, terrified that the wooden supports would give way under the weight. They were well aware of the flimsy construction built by their colleagues over the past year. They were also concerned that Allied bombers might attack the train, as the Japanese had made no effort to mark it as carrying PoWs.

Getting back down to Kanchanaburi hospital camp probably saved Frank's life. Eggs, peanuts and fruits were readily available, providing him with a much-needed source of vitamin B.

The commandant at Kanchanburi was Colonel Toosey of the 135 Field Regiment Royal Artillery, the British colonel that David Lean chose to base his main character Colonel Nicholson on in his film *The Bridge on the River Kwai*. In the film Nicholson (Toosey) is portrayed as a camp commander who helped the Japanese in the building of the railway, but this was far from the truth as he did everything in his power to hinder the Japanese. He quickly realised that if his men did not make an effort to build the bridge, they were likely to be punished by the Japanese, with rations withheld and the subsequent loss of many lives. Toosey cleverly walked a fine line between helping the Japanese build their railway, whilst ensuring that his men were looked after and fed. Not all the men subscribed to Toosey's ideas and many described him as 'Jap Happy', but he was indeed a great leader and saved many men's lives.

I was fortunate to meet Colonel Toosey's son Patrick some years ago and we have since become good friends. Ironically, Patrick's life was saved by Frank's portable defibrillator when he had a heart attack in Liverpool on 23 February 2010 and a second one on 5 May the same year in Lime Street station.

On 16 December 1943, along with 500 survivors of F Force, most in a very bad state, Frank was transferred by train back down to Singapore. It had been eight torturous months since he had left Changi and more than 3,000 of his colleagues in F Force had died.

Conditions on the train this time were slightly better than his journey up some nine months previously. There were only twenty to a truck instead of thirty and some of the wagons had straw bales for the men to lie on. Frank's train stopped at regular intervals to bury some of the less fortunate men who died on the journey and they were always given a decent burial along the track side. Graves were marked by simple wooden crosses, later to be exhumed by the Commonwealth Graves Commission and reburied in cemeteries.

Around ninety-five per cent of those who survived the ordeal were still heavily infected with a variety of tropical diseases and half of them had to be hospitalised on arrival into Changi. The men who had remained in Changi were shocked and stunned at the condition of the new arrivals and did their best to make them as comfortable as possible. One man who had not left Changi commented at the time:

*When we were reunited with F Force, there was no doubt that whilst we had been to some degree deprived of necessary nourishment, the appearance of the survivors from F Force was such that we were absolutely shocked. They all looked like walking cadavers, giving the appearance of skeletons over which a yellowish-green skin, translucent and almost glowing, had been stretched. And these were the healthiest of the survivors. Amazingly they were in good spirits at having arrived back to the comparative luxury of Changi and told us stories of death and survival in the various camps along the railway.*

Still suffering from the effects of beriberi, Frank was immediately put into Roberts hospital, where he was given the best food and treatment available despite the shortage of both food and medicine.

The Royal Victoria Hospital, Belfast circa 1940.

The medical staff at the Royal Victoria Hospital, Belfast in 1939. Frank is sitting front row, end right.

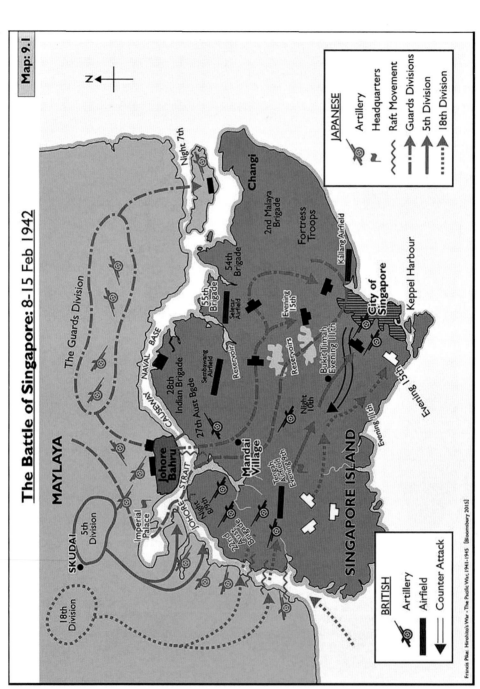

Map of Singapore Island in 1942 showing the Japanese advances.

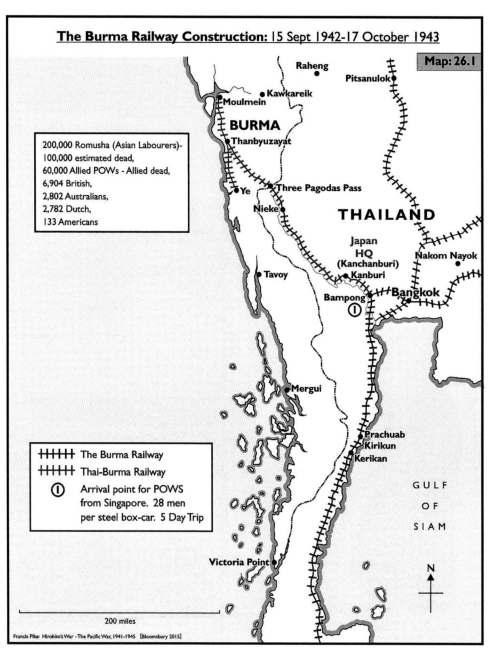

Map showing the trace of the Thai/Burma Railway in 1945 and the statistics of the numbers of Asian labourers and Allied prisoners of war involved.

Frank's Japanese record card. The Japanese kept detailed records of all prisoners of war and their movements during captivity.

Drawing of the Selerang incident in 1942 where Frank, my father and 17,000 of their colleagues were crammed for several weeks.

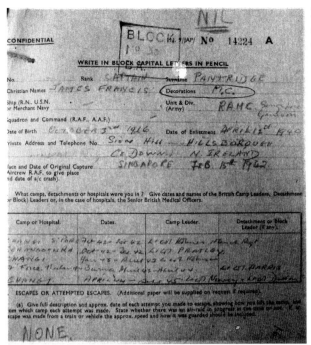

Frank's release questionnaire. Filled out in his own handwriting on return to Southampton in October 1945.

The author and Rod Beattie of the Death Railway Museum, Kanchanaburi sitting on one of the few remaining rail wagons. It was into such wagons that Frank and my father were packed on their way from Singapore to Thailand.

The bridge over the River Kwai at Kanchanaburi. This is not the original bridge built by the prisoners, but a new metal bridge built after the war.

The SS *Almazora*, the ship that carried Frank home to Southampton in October 1945.

An improvised operation in a medical hut on the Thai/Burma Railway under basic conditions.

Frank as a young officer in the Royal Army Medical Corps circa 1940.

The author in the Chunkai cutting near Kanchanaburi, Thailand. The cutting was hewn out by hand using basic equipment.

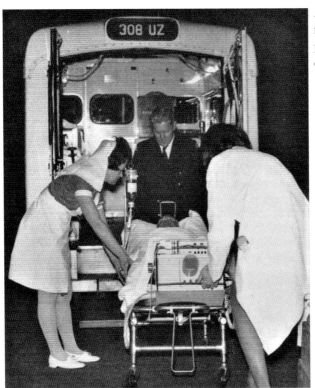

A patient being loaded into the first ambulance that contained a portable defibrillator in Belfast circa 1967.

The original ambulance that held Frank's early portable defibrillator. The vehicle is preserved and housed in the Ulster Transport Museum, Belfast. (*Photo courtesy of the Ulster Transport Museum*)

Resident medical staff of the Royal Victoria Hospital, Belfast in January 1946, three months after Frank's return from the Far East.

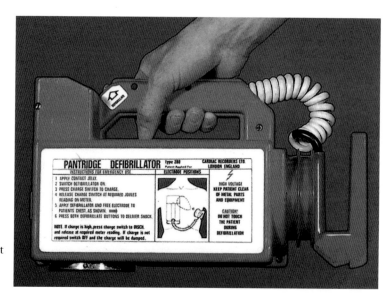

One of Frank's earliest hand-held portable defibrillators.

Frank in studious mood in his office in the RVH.

The granite stone installed in his honour outside the main entrance to Lisburn Civic Centre.

The statue of Frank Pantridge, sculpted by John Sherlock, outside the main entrance to Lisburn Civic Centre.

Frank being presented with a statuette of a gaucho in 1999 by the President of Uruguay in recognition of his work on defibrillation.

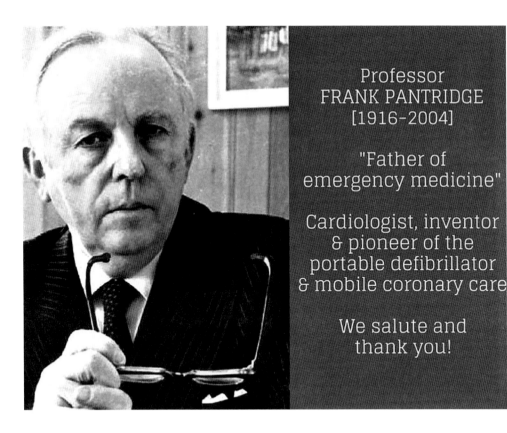

Frank late in life enjoying his pipe and a wee dram.

The cover of Frank's autobiography published in 1989.

The unveiling ceremony of Frank Pantridge's statue, 12 February 2006 by the honourable Timothy Knatchbull, grandson of Lord Louis Mountbatten. Left to right: Alderman Jim Dillon, Dame Mary Peters, The Rt Honourable Timothy Knatchbull, the Mayor of Lisburn Councillor Jonathan Craig and Jack Kyle.

The gravestone for Frank Pantridge in St Malachy's Church, Hillsborough.

He remained in the hospital for six months, before being transferred to Selerang hospital in July 1944. Whilst still in a bad way, Frank of course did his best to help the medical staff treat his colleagues, many of whom were in a worse state than him.

Back in Britain, details about the suffering of the thousands of Far East prisoners of war who had been taken at the fall of Singapore and elsewhere, were sketchy. Other than the postcards home mentioned earlier, the families of the men had little idea of what their loved ones were going through. Very few prisoners escaped from the vice-like grip of their Japanese captors, and those that did and were recaptured, were usually executed.

By mid-1944 however, things began to change. The Japanese had been transferring prisoners in what has been dubbed 'hellships' around the various countries that they had occupied since June 1942. The Americans were now pushing the Japanese back towards their home islands and with submarines operating all over the South China Sea, such transports were very vulnerable to attack. On 12 September 1944, the *Rakuyo Maru*, en route from Singapore to Japan and carrying 1,318 allied prisoners, was sunk by an American submarine in the South China Sea. Most of the prisoners died in the attack or drowned in the sea afterwards, but after drifting around on life rafts for three days, several groups of men were picked up by the *USS Pampanito*. They were the first allied soldiers to be rescued and released from captivity.

It was another six weeks before this group of survivors reached Britain and were able to tell their harrowing stories, the first details to emerge about the Japanese treatment of prisoners of war in the Far East. After medical treatment, these men were debriefed and a full report for the government was prepared. They were strongly advised – no, ordered – not to speak to the press and any of the public about their experiences.

On 17 October 1944, the Secretary of State for War Sir James Grigg, solemnly addressed the House of Commons about the reports of the men:

*Some 150 survivors from a sunk Japanese transport carrying United Kingdom and Australian prisoners of war from Singapore to Japan were rescued by United States naval forces in September. The result of preliminary examinations of the men gives at last a first-hand account of the way our men were treated in the Southern areas of the Far East; and there is no doubt about the policy which was pursued by the Japanese military authorities towards prisoners of war in the areas, which include Burma, Siam (Thailand), Malaya and the east Indies.*

*The prisoners were sent by rail to Siam (Thailand) so crowded into trucks that they could not even lie down during the journey. They were then marched 80 miles. This and subsequent movements in Burma or Siam appear to have been on foot regardless of distance, weather, or the prisoners' state of health. The prisoners were then set to work on the construction of a railway through primitive, disease infested jungle passing over the mountain range between Siam and Burma.*

*The conditions under which these men lived and worked were terrible. Such accommodation as was provided gave little or no protection against tropical rains or blazing sun; worn out clothing was not replaced; soon many lacked clothing, boots and head covering; the only food provided was a pannikin of rice and about half a pint or less of watery stew three times a day. But this work had to go on without respite whatever the cost in human suffering or life. The inevitable result was an appalling death rate, the lowest estimate of deaths being one in five. All the rescued men tell of the amazing way in which the morale of the prisoners has remained high despite the worst the Japanese could do.*

> *In particular, tribute is paid to the medical officers who were captured with them and who have achieved little short of miracles in looking after the sick and injured despite lack of essential medicines, instruments and hospital equipment. There is no doubt that the Japanese military are to blame. To the relatives and friends of all the prisoners concerned, our deepest sympathy goes out.*

It was January 1945, before Frank was deemed well enough to be transferred from Selerang hospital back to Changi. On arrival he was assigned to assist a medical officer in the dysentery ward, '*a most depressing assignment*,' he said. Perhaps he might have actually said that it was a 'shit assignment'.

A few months later he was fortunate to be transferred to one of the general wards where he worked under the wonderful Australian doctor Bill Bye. The two men hit it off immediately and, with little drugs available, they did their best for the men under their care who were suffering from a range of diseases from diphtheria to malaria. Many of them had stopped eating and Frank spent much of his time talking to them and encouraging them to eat.

By early 1945, with Allied bombing raids occurring all over the east, the Japanese were beginning to lose the war. Prisoners all over the Far East became concerned that if the Japanese lost the war, they might all be massacred. Tensions began to rise in the camps and they became much more on edge as the war drew to a close. Whilst Frank was now in Changi, my father was still up on the railway in Chungkai driving supplies around for the Japanese. They feared that the large numbers of prisoners across the camps scattered across the Far East might revolt and murder the limited number of guards in each camp. Consequently, an order was sent out to all camp commanders that if an allied invasion of the Japanese home islands was to occur, all prisoners were to be annihilated and all evidence destroyed.

A copy of the document was found in the American Archives after the war. After translation it reads:

**The Time:**
*Although the basic aim is to act under superior orders, individual disposition may be made in the following circumstances:*

*(1) When an uprising of large numbers cannot be suppressed without the use of firearms.*
*(2) When escapees from the camp may turn into a hostile fighting force.*

**The Methods:**
*(3) Whether they are destroyed individually or in groups, or however it is done, with mass bombing, poisonous smoke, poisons, drowning, decapitation, or what, dispose of them as the situation dictates.*
*(4) In any case it is the aim not to allow the escape of a single one, to annihilate them all, and leave not any traces.*

In Changi the prisoners were all well aware by the beginning of August that the Japanese were losing the war. On 7 August, news filtered through via the 'tweets' of the hidden 'canaries', that a new type of bomb had been dropped on Hiroshima by the Americans, killing thousands of Japanese. There was talk amongst the men of the Japanese surrendering, but no one quite believed it. A few days later, news of a second bomb dropped on Nagasaki came via the earliest form of 'Twitter'; again, the prisoners were sceptical, as they had already had so many false dawns over the past three and a half years. The 'tweets' turned into reality on 14 August 1945 when Japan surrendered unconditionally.

The following day Emperor Hirohito broadcast to his subjects:

*After pondering deeply, the general trends of the world and the actual conditions obtaining in our empire today, we have decided to effect a settlement of the present situation by resorting to an extraordinary measure.*

*We have ordered our Government to communicate to the governments of the United States, Great Britain, China and the Soviet Union that our empire accepts the provisions of their joint declaration.*

*To strive for the common prosperity and happiness of all nations as well as the security and well-being of our subjects is the solemn obligation which has been handed down by our imperial ancestors and which we lay close to the heart.*

*Indeed, we declared war on America and Britain out of our sincere desire to ensure Japan's self-preservation and the stabilisation of south East Asia, it being far from our thought either to infringe upon the sovereignty of other nations or to embark upon territorial aggrandizement.*

*But now the war has lasted for nearly four years, despite the best that has been done by everyone – the gallant fighting of the military and naval forces, the diligence and assiduity of our servants of the state and the devoted service of our 100,000,000 people – the war situation has developed not necessarily to Japan's advantage while the general trends of the world have turned against her interest.*

*Moreover, the enemy has begun to employ a new and most cruel bomb, the power of which to do damage, is indeed incalculable, taking the toll of innocent lives. Should we continue to fight, it would not only result in an ultimate collapse and the obliteration of the Japanese nation, but also it would lead to the total extinction of human civilisation.*

*Such being the case, how are we to save the millions of our subjects, or to atone ourselves before the hallowed spirits of our imperial ancestors? This is the reason why we have ordered the acceptance of the provisions of the joint declaration of the powers.*

*We cannot but express the deepest sense of regret to our allied nations of South East Asia, who have consistently cooperated with the Empire towards the emancipation of East Asia.*

*The thoughts of those officers and men as well as others who have fallen in the fields of battle. Those who have died at their posts of duty, or those who met with death (otherwise) and all their bereaved families, pains our heart night and day.*

*The welfare of the wounded and the war sufferers and of those who have lost their homes and livelihood is the object of our profound solicitude. The hardships and suffering to which our nation is to be subjected hereafter will be certainly great.*

*We are keenly aware of the inmost feelings of all of you, our subjects. However, it is according to the dictates of time and fate that we have resolved to pave the way for a grand peace for all the generations to come by enduring the (unavoidable) and suffering what is insufferable. Having been able to save face and maintain the structure of the Imperial State, we are always with you, our good and loyal subjects, relying upon your sincerity and integrity.*

*Beware most strictly of any outbursts of emotion that may engender needless complications, of any fraternal contention and strife that may create confusion, lead you astray and cause you to lose the confidence of the world.*

*Let the entire nation continue as one family from generation to generation, ever firm in its faith of the imperishableness of its divine land, and mindful of its heavy burdens of responsibilities, and the long road before it. Unite your total strength to be devoted to the construction of the future. Cultivate the ways of rectitude, nobility of spirit, and work with resolution so that you may enhance the innate glory of the Imperial State and keep pace with the progress of the world.*

The Second World War was finally over.

In Singapore, prisoners spilled out of the camps onto the roads shouting and cheering. Hidden Union flags were dug up, Japanese flags were torn down and once again the Union flag flew over British possessions in the Far East.

Almost immediately after the surrender, the RAF and AAF began dropping food by parachute into the camps all over the area and some even dropped men from the three services to help. Eric Lomax, author of the *Railway Man* remembers:

*They looked terribly young to us: a British officer from the air force, one from the navy and one from the army, with a priggish and bossy air about them, coming to take charge of us. We did not feel helpless, or in need of rescue by such inexperienced young men. The captain was told by a PoW that he had been in school when we were first locked up and that if he liked, we would give him lunch, but that was the only co-operation we were going to give him.*

With the war now over, the responsibility for getting the men home lay with an organisation called Repatriation of Allied Prisoners of War and Internees (RAPWI), or 'retain all prisoners of war indefinitely' as the prisoners called it. With men spread over a million square miles, it was an exercise that was to prove extremely difficult and time consuming. Three quarters of a million armed Japanese troops were still active across the Far East and RAPWI was unsure as to how they would react to the Emperor's surrender. The Japanese had been trained and brainwashed into a 'fight to the death' mentality.

Leaflets were dropped from allied aircraft giving them some basic advice:

**To all Allied Prisoners of War**
*The Japanese have surrendered unconditionally, and the war is over. We will get supplies to you as soon as is humanely possible and will make arrangements to get you out but, owing to the distance involved, it may be some time before we can achieve this.*

*YOU will help us and yourselves if you act as follows:*

1. *Stay in your camp until you get further orders from us.*
2. *Start preparing nominal rolls of personnel giving fullest particulars.*
3. *List your most urgent necessities.*
4. *If you have been starved or underfed for long periods DO NOT eat large quantities of solid food, fruit or vegetables at first. It is dangerous to do so. Small quantities at frequent intervals are much safer and will strengthen you far more quickly. For those who are really ill or very weak, fluids such as broth and soup, making use of the water in which rice and other foods have been boiled, are much the best. Gifts of food from the local population should be cooked. We want to get you back home quickly, safe and sound, and we do not want to risk your chances from diarrhoea, dysentery and cholera at this stage.*
5. *Local authorities and/or Allied officers will take charge of your affairs in a very short time. Be guided by their advice.*

Frank of course, as a doctor, was well aware of the dangers of overeating and limited himself to small portions on a regular basis. Despite the warnings however, such advice was ignored by many of the men who had not seen proper food for over three years, and they proceeded to gorge themselves on the supplies that had been dropped. Quite a few were to pay the price for their indulgences, as bloating and stomach upsets stuck with them for some time afterwards. One man was reported to have eaten twelve cooked breakfasts one after the other.

In Singapore, the Japanese officers were devastated by the orders to surrender. It ran against their training in the cult of *Bushido*, where surrender was to bring shame on their families back in Japan. Three hundred officers attended a farewell party in Raffles Hotel where, after several glasses of sake, they committed suicide by either holding a grenade to their stomachs or disembowelling themselves with their swords.

When the surrender came, Admiral of the Fleet Louis Mountbatten, who commanded the British South East Asia Command, was sitting on his ship in the Straits of Malacca. Mountbatten had been waiting for instruction to launch Operation Zipper, a British plan to capture either Port Swettenham or Port Dickson as staging areas for the recapture of Singapore. Concerned about the condition of the thousands of PoWs, he made his fleet ready to steam into Keppel Harbour, Singapore but was prevented from doing so by American General Douglas McArthur, supreme commander in the Far East. McArthur did not want anyone upstaging his formal acceptance of the Japanese surrender by Emperor Hirohito scheduled for 2 September. This two-week gap accounted for the deaths of many PoWs, an action for which many who survived never forgave McArthur.

The men wrongly blamed Mountbatten for not liberating them earlier, giving him the nickname 'linger longer Louis'. Frank was to come across 'linger longer Louis' later in tragic circumstances as we will find out later in the book.

*Chapter 6*

# Homeward Bound

Whilst now reasonably well fed but still a skeleton of a man, Frank sat waiting in Changi. He was totally frustrated and was chomping at the bit to get home to his family and his native Northern Ireland.

On 11 September 1945, the hospital ship M*S Oranje* docked in Keppel harbour Singapore. By co-incidence, on board was a former friend of Frank's, Doctor Tom Milliken, the two men having been students at Queen's University some ten years previously. Milliken had been made aware that his old friend Frank Pantridge was in Changi. He recalls the ship's arrival into Singapore:

*I was medical officer of the* Oranje, *a Dutch luxury liner which had been converted into a hospital ship with 700 beds. In June 1945 we sailed to the Middle East, picked up Australian and New Zealand sick and wounded for repatriation and disembarked them at Freemantle and Wellington. During the return journey the atomic bomb was dropped on Hiroshima and we were diverted to Singapore to pick up FEPOWS.*

*We berthed on Friday 11 September 1945, the day after the cruiser HMS* Sussex. *I commandeered a jeep to get to the PoW camp at Changi where, at the gates I found a mass of Australians. They had heard a rumour that Mountbatten was coming and thought that I was he, presumably because of the white drill uniform and the fact that I was flanked by two army officers of field rank.*

*The Australians were in a happy mood and demanded a speech. It was easy to get them to understand by my accent that I was not Mountbatten and that I was looking for a colleague, Major Pantridge. Ah! They said, 'the doctor'. It was clear that the rough skeletons had acquired a considerable respect for the 'doctor'.*

*I found Frank in one of the many huts. He was trying to put on a shirt but with little success because of his weakened and wasted arms. Frank said, 'Hello Tom', in an offhand almost casual manner as if he met me every day, even though it had been six years since we were students together at Queen's University. The upper half of his body was emaciated skin and bone, the lower half was bloated with the dropsy of beriberi and he weighed under five stones. The most striking thing was that his blue eyes blazed with defiance. He was a physical wreck, but his spirit was obviously unbroken. The eyes said that he was indestructible. On the journey to the* Oranje *Frank hardly spoke. When he got to my cabin he asked for a beer and promptly went to sleep. At dinner later that evening with the ship's captain, he ate steadily throughout the meal without comment.*

The *Oranje* was shortly due to leave for Australia, but there was no way that Frank wanted to go there, he wanted to get back to Northern Ireland as soon as possible: '*I need to get on a ship to England, can you sort that out for me?*' he said to Tom. '*Leave it with me, Frank,*' Milliken replied.

Although he had not been given formal orders to leave Singapore, Milliken arranged for Frank to get on board the *Almanzora*, waiting in Keppel Harbour soon to depart for Southampton. Frank remained indebted to Tom Milliken for the rest of his life and indeed said later, '*I would lay down my life for Tom Milliken.*'

The *Almazora*, one of the first ships out of Singapore after the Japanese surrender, was soon due to sail for Britain with 1,000 PoWs and a group of civilians on board. Ironically it was a ship built by Belfast shipbuilders Harland and Wolf in 1914. On boarding the following day, Frank discovered many of the passengers in a state of revolt due to what they considered to be poor conditions. As he settled down in the hold with his meagre belongings, pandemonium broke loose up on the decks. The civilian passengers were so disgusted with the conditions, that they revolted, walking off the ship in protest and returning to Sime Road camp. The *Straits Times* reported the revolt on 12 September:

> ***Angry Internees Protest – Walk off ship and return to Sime Road camp***
> *Civilian internees liberated only 10 days ago from Sime Road camp, Singapore are so disgusted and disillusioned with the conditions under which they are expected to sail home to the UK, that some of them have walked off the ship, where they were living in mess decks, and have returned to Sime Road.*
>
> *The internees complaints about conditions on the repatriation ship are set out in a letter to the Editor of the* Straits Times *by Mr L.D. Whitfield of the Federated Straits Education Department, who spent three years in Sime Road camp and is now on board the SS Almanzora. Mr Whitfield writes: 'Our main complaints fall under two heads, accommodation and feeding. Feeding: For supper we were given a small hunk of bread and a small portion of butter. Less than 2 ounces of cold beef and two ounces of pickled cauliflower about the size of a walnut and mug of tea with milk and sugar. We were given nothing more until 8am the next day when we were given the same inadequate quantities of bread, four spoonfuls of porridge and some battered egg. Accommodation: we were between decks strung up in hammocks where we could hardly move, latrines and washing arrangements are disgraceful.*
>
> *A number have already left the ship and returned to camp, but now an order has been given prohibiting us from leaving the ship.*

The ship was called a 'hell ship' by many of the civilian passengers due to the overcrowding and poor food, but of course Frank and his fellow PoWs were just happy to be on their way home. Conditions, whilst basic, were a whole lot better than they had endured during the previous three and a half years as prisoners of war.

The *Almanzora* departed Keppel Harbour on 17 September and as the ship steamed up through the Straits of Malacca, Frank gazed back towards the coast of Malaya and to Thailand further north. He could not help but remember the thousands of his colleagues who lay buried there. They too had hoped and prayed for release from captivity as much as he had, they too had dreamed of happy reunions with their families, but whoever controls the destiny of man had ruled differently, their bodies lay instead in unknown graves in a foreign land.

The *Almanzora's* first port of call was at Colombo Ceylon, five days later. As it docked, lights flashed, rockets split the sky and sirens blared welcoming the returning prisoners of war. A group of Women's Royal Naval Service (WRENS) waved furiously as the ship edged into the harbour.

From Colombo the ship sailed on to Trincomalee, a port city on the north-east coast of the island and into Taufik at the southern end of the Suez Canal. From there, it steamed through the Suez Canal and into the Mediterranean, where Frank raised a glass or two with his colleagues to celebrate his twenty-ninth birthday on 3 October. After a brief stop in Gibraltar, they eventually arrived into Southampton on 17 October 1945, a journey that had taken exactly a month.

The national press at the time reported the homecomings of the returning PoWs in a fleet of ships, with the *Daily Telegraph* reporting:

> *Liners with nearly 20,000 former prisoners of the Japanese aboard are racing home to Britain from ports in South-East Asia, it was announced yesterday. Among the many famous liners acting as merchant ships are* the Empress of Australia, *the* Chitral, *the* Ormonde, *the* Almanzora *and the* Worcestershire.

92  *Frank Pantridge*

My father came home to Southampton on the *SS Ormonde*.

When the *Almanzora* docked in Southampton, the returning PoWs were greeted by the Mayor of Southampton J.C. Dyas, the Chief Constable of the city and General John Crocker. Crocker had been the General Officer Commanding I Corps during the D-Day landings on 6 June 1944.

The mayor welcomed them over the ship's loudspeakers: 'I offer you all the town's sincere thanks and gratitude for your services to the country and I hope that what you have endured will always be remembered.'

Frank was unimpressed by the fawning speeches and just wanted to get home to his beloved Northern Ireland.

At the foot of the gangway they were all given a copy of the King's Message, along with a document entitled 'Guard your tongue':

*You are news now and anything you say in public or to the press is liable to be published throughout the world. You have direct or indirect knowledge of the fate of many of your comrades who died in enemy hands as a result of brutality or neglect. Your story, if published in the more lurid or sensational press, will cause much unnecessary unhappiness to relatives and friends.*

*If you had not been lucky enough to have survived and had died an unpleasant death at the hands of the Japanese, you would not have wished your family and friends to have been harrowed by lurid details of your death in the sensational press. That is just what will happen to the families of your comrades who did die in that way if you start talking to all and sundry about your experience. It is felt certain that now you know the reason for this order you will take pains to spare the feelings of others. Arrangements have been made for you to tell your story to interrogating officers who will then ask you to write it down. You are not to say anything to anyone until after you have written out your statement and handed it in.*

They were then loaded into coaches and driven to a transit camp on Southampton Common. The route to the Common was lined with waving and cheering people giving the men a warm welcome back to Britain.

In the transit camp Frank was delighted to find another Northern Ireland man, a planter from Malaya called John Spiers, an officer in the Malay Volunteer Force (MVF). The two men found the camp a dreadfully boring place and using their Irish wit and charm they made a quick getaway the day after they arrived. The duo went AWOL and caught a train to London and from there on to Liverpool where they caught a ferry to Belfast, or 'God's own country' as they put it.

*Chapter 7*

# Belfast and the RVH again

Tears might have welled up in Frank Pantridge's eyes when the Liverpool to Belfast ferry entered the mouth of Belfast Lough and made its way towards the docks. As he stood on the foredeck, probably ignoring the cold and the brisk westerly wind that was whipping the tops of the waves, his mind might have drifted back to the young men lying in far eastern lands who would never have the pleasure of seeing their homeland or loved ones again. As the ship edged into the dockside there was no welcoming party waiting as there had been in Southampton. It was eerily quiet, with just a few groups of people waiting to greet friends and relatives. Frank makes no mention in his autobiography about his homecoming, nor can his nephew remember him speaking about it. It is possible though that they were not aware of his arrival.

Stepping off the ship at Belfast's Victoria terminal, he would have been dismayed at the state of his home city. The previous five years of war had left the citizens reeling. With typical Ulster stoicism however, they were forging ahead to turn it back into its former glory and it is not hard to imagine that as he gazed across to the Cave Hill, he had a burning desire to be part of that process.

On his return from the Far East, Frank was distrustful of politicians for the rest of his life. He blamed Winston Churchill and the Allied officers for the disastrous Malayan campaign resulting in his three and half years of hell.

Despite still being very weak from the weight loss, vitamin deficiency and from hypertension, Frank's first priority was to resume his medical career. Whilst many returning prisoners of war were still bearing the scars of their imprisonment, they had to face reality: the world had not stood still and despite their experiences they had to get on with their lives and this is exactly what he did.

With six months left to complete before he would be fully qualified as a doctor, Frank managed to get a houseman's position in the RVH outpatients department at a salary of £1 per week, plus free board and lodgings. There, he reported to the senior physician R.S. Allison. Allison had been in the navy during the war and still behaved as though he was in command of a battleship rather than a hospital department.

Whilst his ambition was to become a GP, on completion of his final six months no GP positions were available in or around Belfast, so he had to look elsewhere for something challenging. The only position available at the time was as a part-time lecturer in pathology at QUB, pay being the small sum of £150 per annum. It would appear that he was not enamoured with his work in pathology, when he said: *'There was nothing one could do for the individual who had already reached the mortuary and if the cause of death was not already known, it was seldom established there.'*

Having suffered from beriberi in Japanese prisoner of war camps, Frank began to devote his attention to the effects of it on the heart. Caused by a vitamin B-1 deficiency, also known as thiamine deficiency, wet beriberi affects the heart and circulatory system. In extreme cases it can cause heart failure. A year after his return, he was still suffering from the disease and the memories of his colleagues who had died from failure of the heart due to its effects remained vivid.

His first project was to undertake a study of the effect beriberi had on the heart. As animals were used for medical research at that time, he chose pigs. A pig's heart resembles that of a human more closely than any other animal, so it was a logical choice. In 1946, the Ministry of Home Affairs in the United Kingdom were opposed to vivisection and to carry it out you needed a licence. Such trivia never worried Frank and he carried on his research without even thinking about obtaining a licence. There was a devilish streak in him which delighted in breaking the rules, giving the two fingers to authority and defying the establishment.

Of course, with the family owning a farm, he was able to purloin a pig quite easily and when he brought it into the RVH it caused a great deal of concern and comment from the rest of the staff. (It's worth noting at this point, that over forty years later similar concerns were raised when pig's hearts were experimented on as substitutes for human hearts.)

He also undertook a study of his friends in the camps resulting in a nine-page account of the effect of beriberi on former Japanese prisoners of war.

In 1947, Frank decided that to further his medical career he had to become a member of the Royal College of Physicians (RCP) and in September of that year he set off for London to sit their examinations. '*More in hope than with any certainty of success,*' he said. With his usual degree of luck and determination he passed the exams and was admitted to the RCP.

The National Health Service (NHS) was formed the following July, bringing amazing benefits to the population of the UK. Frank said at the time that: '*Only those who saw medical practice before the 5 July 1948 and subsequently worked in the NHS, could appreciate its tremendous benefits for patients, especially impecunious ones.*'

Shortly after the inception of the NHS, Frank left his native Northern Ireland to take up a fellowship in the University of Michigan, Ann Arbor, west of Detroit. There he worked with Frank Wilson, an expert in electrocardiography.

At that time, Harry S. Truman was still President of the USA, a fact that delighted him as he was always grateful to Truman for dropping the atomic bombs on Hiroshima and Nagasaki. '*President Truman saved my life,*' he said.

After spending a year in the US studying the workings of the heart, he returned to his native Belfast, travelling across the Atlantic onboard a cargo ship along with five other passengers.

Back in Belfast at the RVH, he was appointed registrar under the famous surgeon Sir William Thompson. The two men bonded immediately, as Thompson originated from the village of Annahilt, a mere four miles from the Pantridge family farm. Together the pair set up a unit that performed around 100 mitral valvotomies during the first year. His relationship with Sir William was closely cemented, as his son Doctor Humphrey Barron Thompson had been killed by the Japanese at the battle of Gurun on 15 December 1941. Thompson's son was the medical officer of the East Surreys (my father's regiment) when their battalion headquarters at Gurun was overrun with almost everyone there killed. Around 1830hrs that day, the Japanese surrounded the Surreys' headquarters hut before machine-gunning the flimsy structure at close range. Seven officers lost their lives with only three managing to escape.

Frank had known Thompson's son well as he had also travelled out with him to Singapore on the SS *Strathmore* in March 1941. After my father died in 1990, I found an obituary to Doctor Thompson dated 1945, that had been published in *The Lancet*.

In 1951 Frank's career began to take off when he applied, along with dozens of applicants, for the post of Consultant Physician and Cardiologist at the RVH. He must have impressed the panel as he was appointed to the post. His duties were initially in the outpatient department where he was unable to confine his attention to the workings of the heart.

Two years later in 1953, on the retirement of the senior physician Robert Marshall, he was appointed Physician in charge of the Cardiac Department, one of the youngest people ever to obtain such a position. The first cardiac surgery operations in Belfast, where mitral valvotomy techniques were used, had recently been performed and he went on to publish two papers on this technique, including an evaluation of the operation with valve diseases.

His speciality was coronary arterial disease. Mitral Stenosis is a narrowing of the valve between the atrium and ventricle on the left side of the heart, usually as a result of rheumatic infection. For the next twenty-five years, 2,500 patients had the operation in Belfast, almost all under his care.

Frank always put the welfare of his patients first, going to great lengths to impress this mantra on his staff: '*A patient is not concerned about their chest X-ray, electrocardiogram or electrolyte data. What he or she really wants to know is whether they are going to survive and when they might get out of hospital,*' he would often remind them.

Sometimes a patient would ask him if they were going to die, his stock reply being, '*Yes, and so am I. But I don't know when either of these events is going to happen.*'

Dedicated to his work and with few interests outside of medicine, Frank however, played golf, 'poorly', as he said at the time. One of his colleagues, John Bingham, was captain of the Malone Golf Club on the south west of the city. Bingham invited Frank to play with Fred Daly, who won the British Open in 1947, on his captain's day at Malone. The pair duly won the competition and at the dinner that evening, Daly said that Frank was a genius, '*I've been in the golf game since I was six and I thought that I had seen everything in golf, until I saw him swing a club. With such a swing, only a genius could get the head of the club within a yard of the ball,*' he told an amused audience. Nonetheless, he enjoyed his rounds of golf, a pastime that took him out into the open air and away from his onerous hospital duties. '*I perform much better in the golf club bar than I do on the fairways,*' he said.

Coronary heart disease had reached epidemic proportions in the United Kingdom during the 1950s, and by the early 1960s, coronary care units had been set up in most of the major hospitals. Frank doubted the value of such units, as epidemiological data had shown that the majority of coronary deaths were sudden and occurred outside the hospitals. The death rate of people who died at home of cardiac arrests was seventy per cent. Most of these deaths resulted from ventricular fibrillation, a disturbance of the heart rhythm, where the heart simply quivers instead of pumping blood around the body. This is usually caused by disorganized electrical activity in the ventricles, a type of cardiac arrhythmia. Left untreated, ventricular fibrillation is rapidly fatal as the vital organs of the body are starved of oxygen. This can be corrected by the application of an electric shock of 7,000 volts for around five-thousanth of a second across the chest. His response was '*well we had better go out and pick them* [the patients] *up*'.

Frank was not the first person in Northern Ireland to consider taking a defibrillator out to the people, as several years earlier, Professor Graham Bull of the department of Medicine at QUB thought that it might be possible. Ironically, Frank was highly sceptical of this at the time and said: '*This is yet another of the many idiotic ideas which emanate with monotonous regularly from the Professor of Medicine, who thinks it is possible to achieve immortality for patients with coronary artery disease.*'

He very quickly changed his mind, but it is not known if he apologised to Professor Bull for ridiculing his idea.

In 1963, a survey found that two-thirds of coronary deaths of middle-aged men happened within an hour of an attack, with ten per cent of them happening in the ambulance during the journey to hospital. A year later, in April 1964, when a man collapsed with a heart attack just outside the RVH, he was put on a defibrillator by three doctors almost immediately and survived. This event more than any other, spurred Frank on to develop the portable defibrillator.

Along with John Pemberton and Harry McNeilly of the Department of Social and Preventative Medicine at Queen's

University, he undertook a study that confirmed a theory that patients developing ventricular fibrillation did so during the first five hours from the appearance of symptoms. They said: 'Those who are admitted to hospital represent, in fact, the survivors of a storm which has already taken its main toll.'

In a study in Belfast in 1965 it was discovered that admission to hospital was often delayed by over twelve hours and that out of 998 deaths from coronary heart disease, over 596 died outside hospital with 109 dead on arrival in an ambulance. The average delay between the attack and admission was over eight hours. For Frank, the message was clear: the majority of heart attack victims die unattended at or near the place where they are stricken.

His initial proposals to try and put a defibrillator in an ambulance were met with a frosty and negative response from the general medical establishments and he said at the time: '*We were voices in the wilderness. We were disbelieved and indeed, to some extent, ridiculed. The unfavourable comments emphasised the lack of need for pre-hospital care, the prohibitive costs and the dangers of moving a patient who had had a recent coronary attack.*'

The simplicity of his idea meant that one could instantly grasp it, but it had two technical problems. Firstly, how could one power a standard defibrillator without a mains electricity supply? And secondly, how could medical staff and ambulance personnel move around a piece of equipment weighting over 70 kilograms? (All defibrillators at that time were designed for use in hospitals and were powered by mains electricity.)

Not to be outdone by such problems, Frank, with the help of an engineer, Alfred Mawhinney, developed a unit powered by two car batteries with a static inverter to covert from DC power to 230 volts AC power. During the initial period patients could be defibrillated only inside the ambulance. This was done in the usual way by smearing the patient's chest with a saline-based highly conductive electrode jelly and applying two flat paddles.

With scant regard for authority, the first mobile unit was sent out in a refurbished ambulance of the Karrier Ambulance fleet No.331, and thus the first portable defibrillator went onto the streets of Belfast on 1 January 1966. It was a classic case of Ulster 'string and sealing wax' science at its best. The ambulance carried a trained nurse and a junior doctor. One of Frank's junior doctors, Charlie Wilson was one of the early doctors to go out in the ambulance in 1966. The project was grant-aided to the figure of £2,300 by the British Heart Foundation (BHF) for the first year. The BHF was founded in 1961 by a group of medical professionals, who were concerned about the increasing death rate from cardiovascular disease.

That same year, 1967, Frank wrote several articles in *The Lancet*, articles that caught the eye of five eminent cardiologists in the USA, resulting in them starting to embark on a similar project based on his work.

In May, a meeting in Belfast of the Association of Physicians provided him with an ideal platform for the presentation of the results of the first year's operation of his mobile unit. He told the meeting:

*A signal from a GP was given priority at the hospital telephone switchboard and immediately transmitted to the duty doctor and to ambulance control. The team proceeded with all possible speed to the patient who came under immediate intensive care. The time taken to reach the patient was reduced to fifteen minutes in four out of five cases. In only three per cent of cases was the call for the unit judged to be unjustified. Not one of the 414 patients transported died during transit. This was in marked contrast to the results of a study which showed that 112 of 414 patients with a coronary attack transported by ordinary ambulance died in transit.*

Whilst Frank strongly supported CPR, the evidence backed up his theory that early application of defibrillation was the key to saving lives. He maintained that anyone who was capable of giving CPR was also capable of using a defibrillator and that they should be installed beside every fire extinguisher, as life was more important than property.

The large size and weight of the prototype unit was of concern and in 1968, along with Doctor John Geddes and a bio-engineer Dr John Anderson, he set about the task of making the large clumsy defibrillator in the ambulance into something more portable. They developed a new model, containing a miniature capacitor made for the National Aeronautics and Space Administration (NASA) weighing only 3 kilos. It was put in a bright red box the size of a large transistor radio.

The English cardiology establishment initially showed a lack of enthusiasm for the device, one person even likening his idea to the 'Charge of the Light Brigade', a most unsuccessful venture. Some medical people argued that a defibrillator in the hands of a lay person could be dangerous as their patients might be given a potentially dangerous shock, but Frank was undeterred and got around this by suggesting that it should incorporate a fail-safe mechanism, like the safety catch on a handgun. This would ensure that the unit would not deliver a shock unless lethal arrhythmia ventricular fibrillation was present.

In the USA, reaction to his proposals was totally the opposite to the British establishment views. In September 1967, *Time Magazine* of New York published an article on Frank's units entitled 'Immediate Countershock'. The journalist suggested that:

*Since it was not practical for everyone to have a defibrillator beside him, the alternative was to rush the equipment to the patient. There is just one place in the world where this is being done, Northern Ireland's dour capital city of Belfast. Coronary victims got no benefit from the hospital's well-equipped intensive care unit because one of every four was dead on arrival until Pantridge's portable defibrillator arrived.*

It was suggested that a defibrillator should be installed in the White House as President Lyndon B. Johnson had had a heart attack in 1955 and was deemed to be still vulnerable. This was indeed carried out, with another being installed in his aircraft Air Force 1 soon after. Johnson had cause to be thankful to Frank as he had another heart attack in April 1972 whilst visiting his daughter in Charlottesville, Virginia. He was resuscitated by a mobile coronary care unit using an AED and his life was saved. He credits Pantridge:

*Almost certainly, only one cardiologist has conducted pioneering work that saved my life and those of thousands of others. This distinction goes to Dr James Francis 'Frank' Pantridge, professor at Queen's University, Belfast, Northern Ireland. During a visit to Charlottesville, Virginia, I had a heart attack in 1972, which a mobile coronary care unit successfully treated with a Pantridge defibrillator. I owe my life to the invention of this former Japanese prisoner of war.*

Frank, of course, was delighted that his project had been widely accepted across the Atlantic and his *Lancet* article became a citation classic. It was classed as a highly cited article by the Science Citation Index of Philadelphia. His article had been quoted 260 times over a period of fourteen years.

'*It has been shown for the first time that the correction of cardiac arrest outside hospital is a practical proposition. I am indebted to an American for pointing out that cardiac arrest outside hospital had in fact been corrected for the first time in 1775, when a Danish veterinary surgeon had shocked a chicken into lifelessness, and upon repeating the shock, the bird took off and eluded further experimentation,*' he is quoted as saying, in his usual tongue in cheek manner.

Still the British medical establishment were reluctant to accept the Pantridge Plan, whilst in the US, the Seattle Fire Department started a programme of cardiopulmonary resuscitation. They had a great deal of success and in a random poll of 1,200 people almost 40 per cent of those over the age of twelve had received some form of instruction in cardiopulmonary resuscitation. Over a twelve-year period in that city alone, 1,648 lives had been saved, but although CPR can be important, the earliest possible defibrillation remains the objective of high survival rates as Frank has proved.

By now Frank's reputation and expertise had spread throughout the world and in 1970 he was appointed Canadian Heart Foundation orator.

British Prime Minister Edward Heath appointed William Whitelaw as Secretary of State for Northern Ireland in March 1972 when direct rule of the province from Westminster was imposed. As a leading medic, Frank was invited to dinner one evening in Government House, Hillsborough Castle. The castle was familiar territory for him, lying just across the road from where he was born. The Secretary of State was anxious to find out about Frank's new portable defibrillator and was also anxious to tap into his knowledge of both communities in Northern Ireland. At the end of the evening as he was leaving the dinner, he recounts an amusing story:

*As I took my leave at the end of the evening, I happened to take my pipe, a large one, from my pocket. Whitelaw jumped. I think he imagined I was pulling out a gun. It struck me that a Secretary of State for NI of a less nervous disposition might have been more appropriate.*

A few days later Frank needed to see Whitelaw again in Stormont, Northern Ireland's parliament buildings. British troops had been stationed in the grounds of the RVH and had come under attack from the IRA. Stray bullets had hit the nurses' quarters and the staff were terrified. The IRA were threatening to blow up the hospital if the troops were not removed before midnight and Frank, as usual, put the well-being of his staff to the fore. When this was pointed out to Whitelaw, he became very concerned. '*Do you think the IRA would do such a thing?*' he gasped. '*Certainly not,*' replied Frank. '*I agree with your assessment, but since I have been in Northern Ireland, I have reached the conclusion that neither the inevitable nor the predictable ever happens,*' Whitelaw added. The troops were removed from the grounds of the hospital several hours later.

By 1974, many medical people wondered why, in Belfast, a considerable proportion of patients with a coronary attack got intensive care within one hour whilst in London patients on average, did not notify their doctor within eight hours. On hearing this and being asked '*are Londoners more stoical?*' Frank replied, '*No, more stupid.*'

In 1977 he was interviewed by Tina Brown of the *Sunday Telegraph* who wrote in the following week's paper that: '*She had gone to see the man who could become more important that Christiaan Barnard, the South African who performed the world's first heart transplant.*'

Frank's battle to get recognition for his invention in the rest of the UK continued through the 1970s, but as usual he took on his critics with his usual vigour and proved them wrong. He was awarded a chair in Cardiology by QUB and received an MBE in 1978.

Later that year he was invited to lunch at Buckingham Palace, probably as he said, '*as a result of Tina Brown's article in the* Sunday Telegraph.' Prince Philip was hosting the lunch and there were several knights and a bishop there.

It seems that he had made an impression at Buckingham Palace as the following year on 27 August 1979, a police car screeched to a halt outside his house where he was having a quiet lunch. The driver told him that he had instructions from the Chief Constable to take him across the border to Sligo right away. '*The instruction had come from the Queen herself, sir, as Lord Louis Mountbatten's boat has been blown up and they need a heart specialist urgently,*' the driver gasped.

The car took him to RAF Long Kesh near Lisburn, where a helicopter was waiting to take him to Belleek police station on the Northern Irish side of the border with the Republic of Ireland. When he arrived, there were two other passengers waiting in the helicopter, Lord Mountbatten's grandson and elder brother of Timothy and Nicholas, Norton Knatchbull, and his fiancée. Frank was stunned as to how distraught they were. The helicopter flew them at tree top level and landed in a field covered in cow pats, much to Frank's annoyance as he did not want to walk into Sligo hospital covered in dung. British army helicopters could not cross the border at that time.

At Belleek, they were informed that Lord Mountbatten, one of his twin grandsons Nicholas and Paul Maxwell, a local employed as a boat boy, were dead and another passenger Dowager Baroness-Brabourne, Mountbatten's daughter, was badly injured. Three others who were helping on the boat were also injured, whilst Nicholas's twin brother Timothy had superficial injuries.

They were driven to a bridge across the River Erne which they crossed on foot to be greeted by Garda detectives who drove them to Finner Military camp near Bundoran, from where an Irish army helicopter took them to Sligo hospital. Frank was somewhat concerned as to what reception he might get in Sligo as an Ulsterman.

After he had checked out the injured survivors, he realised that an ophthalmic surgeon was needed as Mountbatten's daughter had serious facial injuries. An urgent message was sent back to the RVH to summon an ophthalmic surgeon. When ophthalmic surgeon Stewart Johnston arrived several hours later, he dealt with the main eye problems much to Frank's relief.

The Garda now wanted to take them back to Classiebawn, Mountbatten's house near Mullaghmore, but Frank said: '*I'm not leaving until I get a stiff whiskey.*' Of course, the press was now swarming all around the hospital and as Frank was leaving, David Capper, the BBC Ireland correspondent knocked on the car window.

Frank wound the window down and said, '*Capper, if you don't move that microphone from beneath my nose forthwith, I'll stick it so far up your arse you will never walk again. Driver, forward.*' Typical Pantridge.

He was invited to spend the night at Classiebawn but he insisted on getting back to his patients and arrived back home in the early hours of the morning where he jumped into a hot bath. Almost immediately his telephone rang, it was one of the Queen's staff from Balmoral Castle wanting to know the condition of the survivors as the Queen was very concerned.

He later received a letter from the Queen thanking him for his efforts, but as he said at the time: '*I felt uncomfortable since I had no contribution to the welfare of the injured, whereas Stewart Johnson most certainly had.*'

It's certainly ironic that Frank was summoned to the scene of one of the most tragic murders carried out by the IRA. Whilst as a prisoner of war, he also knew Lord Mountbatten as 'linger longer Louis' but he certainly respected him. Mountbatten also had a dislike of the Japanese and when Emperor Hirohito made a state visit to Britain in 1971, he initially refused to meet him, eventually relenting under pressure from the Royal family. He did however exclude himself from the state banquet afterwards. Japan was the only country not to be invited to Mountbatten's state funeral.

Seven years later, Frank was to meet Mountbatten's daughter again at the Imperial War Museum during an exhibition of Ronald Searle's drawings done on the Thai/Burma Railway and published in his book *Forty Drawings*. Searle, a fellow FEPOW and an artist, made drawings whilst a prisoner despite the risk and managed to secrete them away until the end of the war. He drew his fellow prisoners and their Japanese guards; he sketched the places and people he glimpsed whilst being moved from camp to camp; he recorded historic moments, the Japanese triumphantly entering Singapore, the planes dropping leaflets that announced the end of the war.

At the exhibition, an emotional Countess Mountbatten told Frank: 'On return to London after the tragedy, my chest X-ray was said to resemble that of a cadaver, but my lungs have returned to normal and my eyes have completely recovered.'

Some years earlier in 1964, Frank bought the house of one of his former lecturers, Edward Mayrs, Professor of Pharmacology at QUB. It was a ramshackle old building close to west Belfast, an area that was a hotbed of unrest during the 'Troubles'. The house was within five miles of the RVH, a regulation that was stipulated by the hospital for its senior medical staff. Frank was able to get to hospital quickly from his house, using a direct route through Andersonstown, a staunchly republican area, but whilst he had been warned about taking such a dangerous route, in true fashion he chose to ignore the warnings. A magistrate friend of his, Bill Staunton, told him that his route to work was 'nothing less than reckless'. (Soon after Staunton was shot and killed by the IRA outside St Dominic's school on the Falls Road.)

Car hijackings were commonplace at that time, and on his journey to and from the hospital Frank was quite often stopped by young men demanding his car: His response was always: '*I'm a doctor in a hurry to get to the hospital to look after your relatives. If you don't get out of my way, the continuity of your anatomy will be in grave danger. They always got the message. My car was never hijacked.*'

During the early 1970s, at the height of the 'Troubles', he was awakened around midnight one night by a policeman knocking on his door. '*I have a message from Sir Jamie Flanagan, the Chief Constable of the Royal Ulster Constabulary (RUC), sir. You are to dress and go immediately to your sister's house.*' said the policeman.

Completely ignoring the instruction, he went to bed and the next morning, phoned Sir Jamie whom he knew well. '*You should make yourself scarce, Frank, as I have information that you are a likely target for terrorists.*'

Of course, being Frank, he again completely ignored Flanagan's advice, even to the extent of refusing a bodyguard. Why he should be a likely target is unclear other than the fact that he was a Protestant.

The 1970s in Northern Ireland were a most difficult time, as I myself can testify, but Frank was never at all phased by the 'Troubles' and he said: '*If I can endure the Malaya campaign and Japanese prisoner of war camps, I can easily cope with a few terrorists.*' A statement that my father also echoed many times during the 'Troubles'.

In 1972 Frank was appointed to the Northern Ireland Council for Postgraduate Medicine (NICPM) under the chairmanship of Sir John Biggart, a man who had little clinical experience and set about cutting back on research fellowships in cardiology. '*There is likely to be little outlook for those who will train in those areas,*' he said. Of course, this angered Frank immensely and at his first meeting of the NICPM, he arrived with all six volumes of Churchill's *The Second World War*. He threatened to read all six volumes at the meeting unless Biggart resigned. The following day Biggart resigned from the committee; Frank had got his way.

During the early 1970s, he continued to lobby the Department for Health and Social Services (DHSS) for the introduction of portable defibrillators in all front-line ambulances across Great Britain and Northern Ireland, but without success. A working party was set up in 1975 to make recommendations on the care of the coronary patient. The party recommended that:

*The DHSS should actively encourage the development of mobile coronary care; the exact means must depend on local conditions, Doctor-manned mobile units should be developed where possible, but in areas where this was not practicable, a service manned by trained ambulancemen or other para-medical personnel should be developed.*

The DHSS refused to accept the committee's advice and Frank became very frustrated.

A DHSS health notice in 1976 declared that:

*Health authorities will wish to know that so far, no firm evidence has emerged that the use of specially equipped ambulances manned by ambulancemen who have received training in advanced techniques significantly affects the overall mortality rate of patients suffering from acute myocardial infarction.*

The Pantridge system was gaining popularity in the US during the 1970s. A front-page article in the *New York Times* of 11 August 1975 read:

*Fifty light-weight defibrillators costing $1,200 each are to be flown to this country today for extensive testing in hospitals, homes, factories, office buildings, streets, stadiums and rural areas.*

In 1980, Frank was invited to take part as chairman of a session at the Asian Pacific Congress of Cardiology in Singapore, ironically the city where he was first incarcerated by the Japanese. For him it was not a sentimental return to his old stomping ground, but more of a clinical experience. Singapore, of course, had altered dramatically over the thirty-five years since he had been there. His old camp in Changi was now an international airport. The junks and sampans were no longer plying their trade up and down the river. The Chinese were now living in high-rise flats and only a few go-downs remained as a tourist attraction. Singapore was a lot cleaner than he remembered it, mainly due to the government of Lee Quan Yu and tourists were flocking to the island.

In 1981, the Medical Editor of *The Times* newspaper wrote:

*Six years after the leading heart specialists recommended that the NHS should provide specially equipped ambulances for heart attacks, the Department of Health has done nothing. When a man collapses with a coronary attack in the streets of Melbourne, Seattle or hundreds of other cities round the world, the passers-by know they can summon an ambulance to give immediate life-saving equipment.*

It was to be another seven years before the DHSS reversed its views and accepted that emergency ambulances across Britain should be suitably equipped with defibrillators and as many as possible of the front-line staff trained in advanced resuscitation techniques. This was subsequently delayed for a further seven years as they dragged their feet; sixteen years after the introduction of mobile coronary care in Belfast, ambulances across the UK were still not equipped with portable defibrillators. According to Frank:

*There were tens of thousands of unnecessary deaths outside hospital from coronary attacks during that time. Coronary care units limited to the hospital were useless because patients, when they got to them by the usual transport, were either convalescent or moribund. I am, of course, glad that the concept of pre-hospital and mobile coronary care has been accepted in the UK, although tragically late and in a somewhat half-hearted way. It is probable that there are still more than 35,000 sudden premature deaths in the UK each year.*

In 1982 Frank's frustrations with the DHSS boiled over and he decided to enter private medicine. It was also around this time that he himself succumbed to illness. One morning he fell whilst getting out of bed and, as luck would have it, he dislodged his telephone off the bedside table and was able to call for help. An ambulance soon picked him up and took him to hospital with a bleeding duodenal ulcer.

The following year, 1981, an American firm tried to sue him for $9.8 million. They claimed that he was in breach of an agreement he had signed with them, and that he had criticised a defibrillator machine they were producing. He had a counter-claim that in fact the firm had broken its agreement with him. He was not in the least bit bothered by such a claim as he had never been money-orientated. Eventually it was settled out of court as he was undergoing a coronary by-pass himself. He was once approached by a man who congratulated him on becoming a multi-millionaire but he had never made any money from it.

In 1986, Richard Needham, Health Minister for Northern Ireland wrote:

*The name of Frank Pantridge is known everywhere in North America, from the humblest ambulanceman in the inner-city ghetto emergency service, to the most erudite members of university faculties in the USA and Canada. Professor Pantridge's name is so well known that he could run for senatorship.*

It was to be 1990, twenty-six years after he had developed the first portable defibrillator, before the then Secretary of State for Health, Kenneth Clarke announced that £30M was to be made available to equip all front-line ambulances in England with defibrillators. How many lives might have been saved during this period had this been implemented sooner, we will never know.

With the success of the portable defibrillator in ambulances assured, Frank turned his attention as to how he could get his defibrillators situated in key locations out in the community where a patient could be given defibrillation as soon as possible after an attack. Whilst his ambulances were saving more lives than ever before, he was aware that people were dying before even the ambulance could get to them. The powers that be in the NHS were concerned that only trained medical personnel should or could use the equipment. Frank however was convinced that his defibrillators could be designed that almost everyone could use them, and he was proved right as they are now commonplace in the community.

Towards the end of his medical career Frank was convinced that defibrillators could be made small enough that they could be carried in a pocket. Whilst such a laudable proposal never materialised, however, miniature defibrillators were developed that could be implanted into the chest of a patient in much the same way as pacemakers have been implemented. As usual the NHS hierarchy in Britain dragged their feet, but the Americans grasped the concept right away. Morton M. Mower, an American cardiologist, set out to miniaturize defibrillators and implant them in the chests of high-risk patients in Baltimore, Maryland. Several cardiology experts doubted the AICD's potential for clinical success. Bernard Lown, the developer of the DC defibrillator, wrote in the journal *Circulation*:

*The very rare patient who has frequent bouts of ventricular fibrillation is best treated in a coronary care unit and is better served by an effective antiarrhythmic program or surgical correction of inadequate coronary blood flow or ventricular malfunction. In fact, the implanted defibrillator system represents an imperfect solution in search of a plausible and practical application.*

The first Automatic Implantable Cardioverter-defibrillator (AICD) was approved for implantation in the United States in 1985. At that time, it was indicated only for patients who had a documented cardiac arrest or life-threatening arrhythmia. The implantation of these devices was difficult and required electrodes to be placed directly on the surface of the heart. The device itself had to be placed in the wall of the upper abdomen as it was too large to be placed in the chest. By the early 1990s however, the devices no longer required wiring directly on the heart, but allowed for wiring that went through the vein to the inside surface of the heart. This made the implant procedure much easier and the recovery much quicker (just overnight).

The technology further improved to allow the devices to be smaller, which allowed implantation of the device in the upper chest, like a pacemaker, rather than in the abdomen. This provided protection against sudden cardiac death while avoiding risks and complications associated with trans-venous leads. It has proved to work particularly well for younger people who have had a sudden cardiac arrest and are at risk of it happening again. With an electrode placed under the skin, the AICD system delivers therapy without the need for wires implanted in the heart, leaving the heart and blood vessels untouched and intact. Frank's invention has now moved on from its original purpose.

Frank Pantridge became so well known in the USA that he was dubbed 'the father of emergency medicine' and it was said that he could have run for political office. In 1999 he was invited as a guest of honour to Uruguay for the 1st Uruguay Congress of Pre-Hospital Coronary care. He said at the time, '*I seem better known in South America that I am in Northern Ireland.*'

His patients always came top of his priority list and he was always trying to get them back home as soon as possible after their attacks. Despite suffering auditory damage from the roar of the artillery in Singapore, Frank was always able to detect diastolic murmurs through his stethoscope to the amazement of his students.

## Chapter 8

# Twilight Years

On his last visit to the US in 1983, Frank gave the Morris Lieberman Memorial lecture to the American Heart Association (AHA) in Baltimore. At the dinner after the lecture at the Baltimore Club he was asked to respond to the toast to the guests given by the chairman. The chairman had droned on for a long time and when his time came, he said to the audience: '*My first speech is my short speech – thank you, and my other long speech is – thank you very much.*' With that he sat down to tumultuous applause.

As a freeman of the Borough of Lisburn, Frank was entitled to 'drive sheep through the borough'. Some of his colleagues said that they wouldn't have put it past him to rustle a few sheep and take up the privilege just for the hell of it. What he actually did was acquire some little plastic sheep and put them in his glasses case on the dashboard of the car.

With more time on his hands after retiring he began to take an interest in local affairs, particularly the political situation in Ulster. People in the province often say: 'if you aren't confused about politics you don't understand,' a statement that he certainly agreed with. He recounts an appointment with a 19-year-old, who as a baby he had treated for narrowing of the valves in the vessel leading to the lungs. The man had applied for a job at Dupont, a chemicals firm in Londonderry and needed a medical certificate. After the examination Frank asked the man how things were in Londonderry. The man replied: '*There is rioting all the time.*'

'*Why do you think that this is?*' asks Frank.

'*We want a united Ireland,*' was his answer.

'*If the rioting continues then Dupont might pull out and you would have no job.*'

'*OK I'll get a job in England then.*'

A baffled Frank said, '*You want a united Ireland even if that means you have to go to England for a job.*'

'*Of course, doctor.*' Frank just shook his head in amazement.

On being quizzed about the situation in Northern Ireland during the 1980s he was adamant that there was a similarity to the situation in Malaya and Singapore prior to 15 February 1942. '*Successive Secretaries of State have shown a great reluctance to risk independent, definitive action,*' he said.

When Emperor Hirohito of Japan died on 7 January 1989, he naturally took a keen interest. Hirohito's funeral took place in Tokyo seven weeks later on 24 February. Several statesmen and world leaders refused to attend, including Bob Hawke the Prime Minister of Australia. Frank fully supported Hawke's stance and said at the time:

*There might well have been fewer world leaders at the funeral had they read a recently published account of the International Military Tribunal for the Far East by the late Arnold C. Brackman. Brackman covered the War Crimes trials in Tokyo as a reporter for the US press. Witnesses told of terrible atrocities, mass murder, the beheading of prisoners, the use of men women and children for bayonet drill, multiple rape and the killing of sick and wounded. And yet today, not only are the names of the war criminals for the most part forgotten, so are their deeds. It is estimated that less than 1 per cent of all Japs against whom there was war crime, were ever brought to trial. The Emperor was not in the dock, as he most certainly should have been and so the hearings were largely a farce.*

Since their return from the Japanese prisoner of war camps, FEPOWs had been campaigning the British and Japanese governments for compensation and an apology. This was turned down by the courts in Japan. It was not until the year 2000, that the then Prime Minister Tony Blair, agreed to a multi-million-pound compensation package for former Japanese prisoners of war in recognition of their 'appalling' experiences. This announcement represented a victory for the veterans. At that time there were 5,654 former Japanese PoWs and 4,663 widows in Britain. After a fifty-year struggle, they were to be paid a one-off payment of £10,000 in recognition of their 'unique experience'.

No one knows if Frank was given or accepted this payment. His nephew is also unaware of him getting the compensation: '*There was nothing relating to it amongst his possessions – no official letters, or anything such like. He was probably wondering why the British Government were paying the money and not the Japanese.*'

Both my mother and father had died by then and neither was able to claim the compensation. It is often said that the British Government dragged their feet on the claim to save money.

# Conclusion

As a prisoner of war, along with my father and thousands of other FEPOWs, Frank Pantridge had been compelled to take a very different mental outlook to that of a free person. For three and a half years, his life had been in the hands of a harsh and brutal enemy, with day to day survival his only thought. Had the Japanese shown even a slight degree of compassion and goodwill, or had they possessed the basic principles of supply and organisation, many young lives could have been saved. In most camps, the Japanese simply ignored the normal rules and regulations of a civilised society regarding the treatment of the sick and injured.

It should be pointed out here, however, that the experiences of prisoners in the hands of the Japanese varied considerably. Depending when and where they were imprisoned, some were better treated and better fed than others, so universal conclusions are impossible to make. There is no doubt, however, that the vast majority of prisoners were inhumanely treated by their captors.

The dropping of the atomic bombs on Hiroshima and Nagasaki in August 1945 ended the war quickly, saving the lives of many thousands of prisoners including Frank and my father. Whilst some historians have asserted that the dropping of the atomic bombs was 'an unprecedented act of barbarism, the responsibility for which devolves on the American and British governments', both Frank and my father were adamant that the bombs saved their lives. I certainly would not be writing this book today if this had not happened. Had the Americans invaded Japan in a conventional way, then many more people may have died.

Conclusion 119

President Harry Truman summed it up in his public statement to the American people shortly after the use of first the bomb on Hiroshima:

*We have used it against those who attacked us without warning at Pearl Harbor, against those who have starved, beaten and executed prisoners of war, against those who have abandoned all pretense of obeying international laws of warfare. We have used it in order to shorten the agony of the war. Nobody is more disturbed over the use of Atomic bombs as I am, but I was greatly disturbed over the unwarranted attack by the Japanese on Pearl Harbor and their murder of our prisoners of war.*

The story of the fall of Malaya and Singapore was made even more difficult for Frank and my father to stomach when it was discovered after the war that Colonel Tsuji, commander of the Japanese forces admitted that his ammunition for an assault on Singapore was almost exhausted. He wrote in his autobiography:

*We had barely a hundred rounds per gun left for our field guns, and less for our heavy guns. With this small ammunition supply it was impossible to keep down enemy fire by counter-battery operations.*

The British media have long been obsessed about the Second World War in Europe such as the lives of prisoners of war in German camps, the Battle of Britain and the D-Day landings. From Far East prisoners of war you will only hear stories about beatings, starvation and being worked to death. Frank Pantridge never forgave the Japanese for their inhumane treatment of him and his fellow prisoners of war and hated them until the day he died. A poignant quote he underlined in a book written by a fellow FEPOW relating to his time as a prisoner of war was:

*My tongue had lost its covering and was red raw. Skin was peeling off the roof of our mouths; lips were skinless and developing scales. All sensitive parts of our anatomy were becoming inflamed, with the scrotum almost raw and often bleeding... the Japs in charge paced up and down waving bamboo sticks or hurling pieces of rock at anyone who did not appear to be busy. Their screaming and menacing attitude became maddening, and although I still contend that mind control enabled many of us to emerge from captivity, there really were times when the mind seemed to become detached from being... we were in the true sense automatons.*

My father also bore a grudge against the Japanese until the day he died. He would never ever buy anything made in Japan, to the extent that he never spoke to me for three weeks when I bought a Japanese car. On his death bed in 1990 at the age of 73, his parting words to me were 'the Japs did this to me'.

If the FEPOWs can take just a crumb of comfort from their experiences, it was the cultivation of a sense of comradeship that remained with them for the rest of their lives. Lord Harewood of the Grenadier Guards, who himself was a prisoner of war at the hands of the Germans, wrote: '*You never really know a man until you have been a prisoner of war with him.*' I know that my father had a close bond with his fellow FEPOWs when he returned to Ireland.

In a speech at London's Imperial War Museum on 5 March 1986, Countess Patricia Mountbatten said:

*The Siam-Burma Railway's survivors had in their grip a thorny, but true, measure-stick against which to place the things that did and did not matter. In Frank Pantridge, who I know from the terrible circumstances of my husband's murder, you have one such person who epitomises such a measure-stick.*

Frank never married and his social life after returning from the camps centred around friends and acquaintances for drinks after work. In his younger days he was darkly good looking with a mystique common in many medical men who day by day enact the drama of life and death. He never lacked female company but as he said himself '*only became excited by wine*'. On several occasions it is said that he proposed to four women on the same evening, but when the effects of the wine had worn off it was work not women which engrossed him.

It was during his drinking sessions that he came up with some of his brightest ideas. He was a genius but with an incredibly forceful nature who lived for his profession.

All over the world today automatic defibrillators are ubiquitous, most of the major airlines have them on their aircraft with staff trained to use them, schools and universities have them along with most factories.

Lisburn City Council gave Frank freedom of the city and erected a statue of him, sculpted by John Sherlock, outside their offices at the Lagan Valley Island centre – possibly the only statue to be erected of a cardiologist in the world. They also named a new underpass on the A1 at Hillsborough after him – The Pantridge Link. In Northern Ireland, he is celebrated as a local hero, giving him household name status and rated as the fourth most popular figure in Ulster behind George Best, republican hunger striker Bobby Sands and Dr Ian Paisley. I myself might just add Lady Mary Peters to that list.

Frank never made any money out of his invention as he never patented it, but he was never motivated by money, only trying to help his fellow man. His pioneering work on pre-coronary care, serves as wonderful example of the right man in the right place at the right time. His military career enabled him to promulgate a very simple idea however, it took many years for him to get under the skin of the London cardiologists but with persistence and dedication, he did just that.

One of his colleagues at the RVH is quoted saying:

> *He had a veneer of arrogance at times, but this often concealed an innate shyness. On a good day Frank looked as though he owned the*

*world, on a bad day he looked as though he didn't care who owned it. He appreciated being stood up to, but this, in truth, was often hard to do. He was a man with staccato delivery and a quickness of wit, which often bore the Irish element of visual humour; the clicked fingers acted as punctuation and the stiff index finger directed the verbal missile: a nattily dressed young consultant resembled 'a walking Christmas tree', an ex-rugby playing surgeon 'a great slab of mesoderm', dermatologists and radiologists were 'scratch' and 'shadow' doctors, respectively.*

Frank Pantridge was certainly a genius, but he could haggle over details almost endlessly. He did not marry and much of his social life was centred around meeting friends and acquaintances over a drink after work. This *'lubrication of the synapses'* as he called it, often seemed dramatically to unleash his academic ingenuity.

His love of a drink occasionally got him into trouble, even though as he said: '*I have an exceptional tolerance for ethanol,*' as he called alcohol. He tells of one incident where he was stopped by the police on his way home from a boozy dinner and breathalysed. He demanded to see the calibration curve on the instrument to ascertain its accuracy. The police were amazed, they had never been asked this before. As he was unable to give a urine sample, they took him to the station for a blood test where he insisted that they take it from an artery as his veins were in bad shape. Frank insisted that a surgeon colleague in the hospital come and take the sample, but as he was not there and so much time had elapsed, the police gave up.

To settle a lengthy negotiation over clinical space he was once heard to say that '*he would accept half, provided he got the bigger half.*' He also had a great dislike of the legal system and avoided subpoenas with great cunning. He did, however, have to appear in a Belfast court as an expert witness on one occasion in 1965, when an insurance company challenged a claim by the wife of Sir Richard Dunbar, the head of the Northern Ireland Civil Service who had been killed in a motor-cycle accident. With his usual straight talking, he helped her win her case. He considered lawyers as a particularly arrogant breed. He believed that legal procedures were tedious, tiresome and time-wasting. During his first few weeks at the RVH as a casualty officer, he had been required to wait around for three whole days in a court room wasting valuable time when he could have been treating patients.

In December 2008 a committee was formed in Queen's University to raise money for the commissioning of an oil painting of Frank to be hung in the great hall. The portrait, by Belfast artist Martin Wedge, was unveiled by Irish rugby legend Jack Kyle. Kyle's mother had been treated by Frank some years previously. The following year, in June 2009, the Department of Epidemiology and Public Health at Queen's University held a two-day conference in Frank's honour.

One of the strangest things for me is that fact that after he left school in 1935, Frank never mentions his parents or his brother or sister or his nephew again in his book *An Unquiet Life*. Quite why this should be I just don't know. Even when he returned to Belfast in late 1945 there is no mention of a family reunion. Most FEPOWs on return, talked at length about the wonderful feeling to be in the arms of their family once again. Frank's nephew was even unsure of his relationship with his father prior to his death, he says, *'It would appear that he didn't have a great rapport with his father, because he indicated to me that his father served as a role model only in the negative sense, as more of an example of what not to emulate.'* However, according to Alun Evans: *'He was very close to his mother and when she died, he said that her death affected him more than the deaths of many of his colleagues during the battle for Malaya and in the prisoner of war camps.'*

Field Marshal Lord Wavell, who was Commander in Chief of Allied Forces in the Far East at the time of the Japanese invasion said in his memoirs:

*War is hateful, war is horrible, but war sometimes brings understanding and progress. If we can only keep alive after victory the spirit and energy that war brings to a virile nation and can direct it to the ends of peace and prosperity, we can do great things.*

Frank Pantridge was the epiphany of such a statement as he certainly did great things on return from a dreadful war. He had a notice hanging on his office wall;

> *People can be divided into three groups*
> *Those who make things happen*
> *Those who watch things happen and*
> *Those who wonder what happened*

There is no doubt that Frank Pantridge belongs to the first group.

The character of Frank Pantridge has, I trust, been so broadened in this book, that those of you who have taken the time to read it, feel well acquainted with him. Frank was a man, whose talents and achievements were so extraordinary, that the further we consider his character, the more he will be remembered by the present generation with admiration and reverence.

This book has told the man's story of incredible bravery in the face of overwhelming odds during the Second World War and his dedication to the concept of the portable defibrillator on his return. He was part of the 'forgotten army' and more than 70 years on, with very few still alive, it is a privilege to set the record straight and to honour him.

In 2002, two years before his death, Frank is quoted as saying:

*If I had known I was going to live to this age, I would have looked after myself better.*

Professor James Francis Pantridge CBE MC MD DSc DMEDSc
FRCP FRCPI FACC
Born 3 October 1916. Died 26 December 2004, aged 88.

*Appendix I*

# James Frank Pantridge – Obituary by Professor Alun Evans

Why do people read obituaries? I suspect for two main reasons: to find out about the lives of the great and to find out about their own lives. Although Frank was great, readers tending to the latter category should persist as Frank had a few imperfections. For example, he did not suffer fools gladly and in Frank's eyes there were a great many of us around. When Frank said 'with all due respect' you knew this was the last thing he meant. Slow-learning young doctors in his ward would feel his hyper-extended index finger jabbing into their sterna with the exhortation to 'get out of your coma'; or he would tell them to go and write up all the cases of cardiogenic shock who had survived, telling them 'I will make you world famous', safe in the knowledge that there weren't any survivors. His most severe humiliation, however, was reserved for medical students who were deaf to heart murmurs. He would ask to see their stethoscopes and examine them most fastidiously, trying them out himself, before announcing that 'The problem lies here,' as he pointed to between the ear-pieces.

Frank's reputation route-marched ahead of him: years before I met him, I was asked by my barber if I had met Frank. He then recounted how he had been cutting a patient's hair in Frank's ward and had looked up to see Frank gesticulating in his direction. The barber bravely walked over to Frank and asked, '*Sir, am I annoying you*'? '*No*', replies Frank in a measured way, and then glowering around at his entourage, added '*and come to think of it, you are the only person here who isn't.*'

Frank was born into the farming tradition in the plantation town of Hillsborough, County Down. In his eminently enjoyable autobiography *An Unquiet Life* he tellingly applies the word 'farouche' to himself. He attended Friends School in Lisburn, County Down, and contrary to another obituary, he was not 'expelled from several schools'. He subsequently read medicine at Queen's University Belfast and, having survived diphtheria, qualified as a doctor in 1939. The day that the Second World War broke out he, Thomas (Toffie) Field and three others joined the Royal Army Medical Corps. The five of them went off to war in bespoke uniforms provided by Toffie's father, who owned a clothing company. Toffie was posted to North Africa and Frank was destined for the Far East.

Frank was seconded to the 2nd Battalion the Gordon Highlanders and for a considerable period a phoney war ensued. The British, convinced that only a seaward attack was conceivable, believed that Singapore was impregnable – thanks to the huge guns which were resolutely trained out to sea. Things began to go badly awry, with the loss of the *Repulse* and the *Prince of Wales* in late 1941, due to a total absence of air cover and the onset of the Japanese landward assault down the Malay Peninsula.

For Frank, the gruesome helter-skelter of war began in deadly earnest. In his autobiography he records that 'One Indian officer came in with a hand on either side of his head. He explained that his head was about to fall off. A massive gash from a Japanese sword had severed all the muscles of the back of the neck down to his spine.'

Towards the end of the Malay Peninsula campaign Frank won an immediate award of the Military Cross in the field for working *'unceasingly under the most adverse conditions of continuous bombing and shelling and was an inspiring example to all. He was absolutely cool under the heaviest fire and completely regardless of his own personal safety.'* Characteristically, Frank makes no mention of these events in his autobiography.

Following the humiliating surrender of Singapore, Frank was taken prisoner of war early in 1942. When Toffie Field learnt of this he is said to have commented, 'God help the bloody Japanese.' In reality Singapore and Malaya had been abandoned. Frank was taken to Changi camp where the treatment dished out was barbaric. One day, on the docks in Singapore the Japanese guards were brutalising their British prisoners when Frank saw some German sailors sunbathing on a surface U-boat. The sailors swam ashore and proceeded to beat up the Japanese guards.

In the spring of 1943, 7,000 British and Australians were taken to build the 1,200 mile² long Siam-Burma Railway. It may be added that the first use to which the railway was put was to ferry prostitutes to entertain Japanese soldiers. The horror of what unfolded is captured to some extent in Frank's autobiography, although, for him, the full reality may thankfully have been dulled due to severe nutritional deficiency. Frank meticulously boiled all his drinking water to avoid cholera, thereby exacerbating the nutritional disease that he succumbed to; cardiac beriberi. Frank survived the 'death camp' at Thanbaya in Burma and, altogether, 90,000 of his comrades perished.

Over two years later in 1945 he was liberated by a colleague from the Royal Victoria Hospital, Tom Milliken, who described him thus: 'The upper half of his body was emaciated skin and bone, the lower half was bloated with the dropsy of beriberi. The most striking thing were (sic) the blue eyes that blazed with defiance' – at this time he weighed less than five stone.

The first thing that Frank did was to dig up his uniform, which he had buried for safe keeping and put it on; Tom said he resembled a scarecrow. His defiance was certainly directed at his former captors but, it must be understood, that it was also directed at the

---

2. This figure is inaccurate. The Railway was actually 380 miles long from Ban Pong in Thailand to Thanbyuzayat in Burma (now Myanmar)

authoritarian high-handedness which had so badly blundered in the Far East. Frank's contempt for petty authority ever afterwards must be viewed in the light of his experiences.

Frank returned to England, and with a friend, in the absence of a welcoming party (for after all he had been a player in a losing theatre of war) he absconded for his beloved County Down. With the war over, Frank toyed with a career in general practice but opted for studying beriberi in pigs. After he passed his MRCP he took up a two-year fellowship in the United States, when he worked with Frank N. Wilson, the then world authority on electrocardiography. He also developed a sound knowledge of electronics, which was to later serve him so well. He also saw Charles Bailey demonstrate his new technique of mitral valvotomy.

In 1951 he was appointed consultant physician in the Royal Victoria Hospital, Belfast. This was, in fact, an outpatient physician post, which was, in time, to evolve into a cardiology post. He set about building up a good working relationship with the surgeons and, by the mid-1960s, 2,500 mitral valvotomies had been performed and other residual problems, such as constrictive pericarditis, were mopped up. Frank would dismiss this as routine, but it brought huge benefits to those who received the operations.

In the early to mid-1960s, the concept of the coronary care unit came to be established in Westernised countries. Frank was aware of epidemiological studies in America that showed that mortality in myocardial infarction was highest in the very early hours, i.e. a significant proportion of death was sudden. In 1965, after discussion with Frank, John Pemberton and Harry McNeilly of the Department of Social and Preventative Medicine at the university began a study of survival in fatal cases of infarction over a one-year period. It confirmed that most patients developing primary ventricular fibrillation did so during the first five hours after the onset of symptoms, so that, '*Those who are admitted to hospital represent, in fact, the survivors of a storm which has already taken its main toll.*'

There is an apocryphal story about Frank having lost a golf ball in gorse and setting fire to the bush. Certainly, he loved to quote the American gangster Willie Sutton, who on being asked why he always robbed banks, replied, 'That's where the money is.' So, it was with Frank's great achievements: having identified a worthwhile goal he would pursue it relentlessly. His response to the high mortality from myocardial infarction was to deliver coronary care to the patient in the workplace, street or home. Cardiopulmonary resuscitation had been shown to be effective and defibrillation was a reality, but the equipment was only portable in the sense that you were a contestant in the world's strongest man competition.

Frank's approach was to commandeer an ambulance and, with the assistance of Dr John Geddes, fitted it out with the help of a British Heart Foundation grant. It went on the road on 1 January 1966. The defibrillator was powered by car batteries, weighing 70kg. Mobile coronary/intensive care was now effectively delivered to the patient at the right time. The epidemic of coronary heart disease was still on the increase in the province and was not to decline until the early 1980s. Frank's innovations continued: the introduction of the first truly portable defibrillator in 1968, weighing roughly the same as a new-born baby, and the concept of the automatic defibrillator which Frank first conceived on Saturday, 6 March 1976 on a train travelling between Ghent and Amsterdam.

Mobile coronary care was avidly accepted in North America and taken up by Grace in New York, Crampton in Charlottesville, Virginia, and Cobb in Seattle. In Great Britain, however, attitudes were more phlegmatic. At a symposium in London in 1974, the distinguished cardiologist Celia Oakley asked how it could be that in Belfast a considerable proportion of patients with a coronary attack got intensive care within one hour while in London patients took a lot longer. Were the Londoners more stoical than the Belfast patients? '*No, ma'am,*' snapped Frank, '*more stupid*'.

Today, Frank's ideas are mainstream medicine throughout North America and beyond. It has been recorded that the name Pantridge is known everywhere in North America from the humblest ambulance of the inner-city ghetto emergency service, to the most erudite members of university facilities. Frank was so well known there that an American colleague commented, 'Literally he could run for political office.' Indeed, in the Americas he is known as the 'father of emergency medicine'. He was elected to the fellowship of the American College of Cardiology in 1974 and in 1999 the combined Latin-American and First Uruguayan Congress on Pre-Hospital Coronary Care invited him as a guest of honour, along with John Geddes. Frank was presented with a statute of a gaucho by the president of Uruguay. I received a postcard of Montevideo on which Frank has written, '*I seem better known in South America that I am in North Ireland.*' This was just one of a long series of awards he received, mainly from outside the United Kingdom.

'Does mobile coronary care work?' A study carried out by Charlie Wilson of Ballymena and Clive Russell of Omagh in the early 1980s strongly suggests that it does. Frank's goal was the early stabilisation of rhythm not only through defibrillation but also by the correction of slow heart rates, which are conducive to ventricular fibrillation. In fact, mobile coronary/intensive care can be justified in humanitarian terms alone: good doctors should ensure that patients have pain relief in the shorted possible time – which may make a modest contribution to stabilising heart rhythms. The importance of early treatment in the acute coronary attack has been increasingly accepted with the advent of thrombolytic therapy.

In Britain, automatic defibrillators are now ubiquitous: for example, British Airways has installed them on all its planes and had trained staff, albeit years after American Airlines and Qantas did so. To mark the millennium and its own silver jubilee, the European Congress of Cardiology, in Amsterdam, held a special exhibition in which Frank was honoured as one of the great contributors to cardiology since 1950.

## James Frank Pantridge – Obituary by Professor Alun Evans

In Frank Pantridge we have lost a pillar whose contribution in so many important areas has been immense. He had an extremely interesting and eventful life. Apart from his adventures in the Far East, he eluded a kidnap attempt by a paramilitary group, and was rushed to Sligo to advise on possible treatment after the horrifying attack on Lord Mountbatten's yacht. He was a man of huge vision and also one of breadth of intellect. He was a keen reader, a keen salmon fisherman, had great taste in architecture, furniture and gardens, was very widely read and an avid follower of cricket. He was a connoisseur of malt whiskey which was old enough to vote, and which he referred to as a 'synapse of medicine relaxant'. Not long ago in a good restaurant he asked for a not particularly obscure malt, which the shamefaced barman did not have in stock. In reparation the barman picked up a large tray loaded with bottles of malt and proffered it to Frank, who quipped, '*I couldn't possibly drink all that.*'

He had a veneer of arrogance at times, but this often concealed an innate shyness. On a good day Frank looked as though he owned the world, on a bad day he looked as though he didn't care who owned it. He appreciated being stood up to, but this, in truth, was often hard to do. He was a man with staccato delivery and a quickness of wit, which often bore the Irish element of visual humour; the clicked fingers acted as punctuation and the stiff index finger directed the verbal missile: a nattily dressed young consultant resembled 'a walking Christmas tree,' an ex-rugby playing surgeon 'a great slab of mesoderm,' dermatologists and radiologists were 'scratch' and 'shadow' doctors respectively.

Frank had an eye for detecting nonsense; after George Bull, the professor of medicine, had introduced the concept of multiple choice questions at a clinical meeting, a group of physicians were walking up the corridor arguing about which of the five answers were correct, Frank came up behind them and observed, '*There is a sixth possibility – the professor of medicine is a bloody fool.*' It must be admitted here that George Bull may have given the idea of pre-hospital coronary care to Frank who was initially hostile to it. Nonetheless it was Frank who developed the idea and invented the portable defibrillator.

Frank often adopted an ultra-right-wing stance, sometimes specifically to infuriate those in whom he had detected a streak of liberalism (which he could spot at forty paces). It came as no surprise that during a strike of cooks at the RVH in 1981, he was prepared to organise the catering himself. This was not a case of strike busting but a desire to ensure that patients did not suffer. His patients, particularly children, loved him. He strongly approved of the original ethos of the NHS and deplored the 'Laconian' bureaucracy that developed.

Although in his early days, Frank was not too concerned about cardiovascular risk factors, latterly, having endured quadruple coronary artery bypass grafting he became a disciple. He eschewed both his pipe and saturated fat, and *An Unquiet Life* contains these memorable lines: '*Common sense, however, would suggest it might be unwise to consume large quantities of milk or milk products. Man is the only mammal which consistently consumes the milk of another species.*'

Frank had learnt which side of his bread not to butter. Until now, he was an incredible survivor: as he remarked recently, '*If I'd known I was going to live to this age, I would have looked after myself better.*' He would reflect on how he had out-survived his surgeon. Even in his decline his wit never deserted him: a nurse unwisely asked him why he fell down and got the swift and immaculately timed reply, '*The reason I fall down is because I can't stand up.*'

As we have seen honours were more likely to come from abroad than from home – although he was made a freeman of Lisburn, County Down, and there is a road named after him in west Belfast. Admittedly, Queen's University Belfast, did include him in its millennium exhibition, 'Faces of Queens', and the following year finally awarded him an honorary degree. Sporadically, correspondents would write to our local newspaper from the United States asking why Frank had not been honoured more fully. Some of us did try to remedy this but to no avail, but perhaps this says more about the honours system than it does about Frank. Word did filter down from above that Frank was viewed as 'controversial'.

In 2001 there was a piece in the *British Medical Journal* entitled, 'a paper that saved my life.' The author was a Sydney based general practitioner whose life had been saved by mobile intensive care units serving the population of four million people; undoubtedly countless others around the world are still alive today thanks to Frank's 'controversial' ideas. Maybe, Frank's view of the world was just too large and uncomfortable to fit into the cosy milieu that is Belfast medicine. Some time ago I mentioned to him that the gene for hibernation had been detected in primates for the first time, suggesting that humans would also carry it, and if it could be activated it would be useful in space travel. In a manner reminiscent of the Duke of Wellington's advice on controlling the sparrow population at the Great Exhibition (sparrow hawks), Frank spat out one word 'surgery'. As another great Irishman Jonathon Swift, wrote, 'when true genius appears in the world you may know him by this sign, that the dunces are all in confederacy against him.'

Why do people write obituaries? In this case because Frank paid me the honour of asking me to. Frank fell out with most of us, but I was lucky enough to have fallen in with him again. In common with many, I miss him hugely. As Frank often said, 'good people are scarce'. As Frank's nephew added recently, 'They are now even scarcer.'

Professor Alun Evans, Queen's University Belfast,<br>December 2004

*Appendix II*

# People Who Knew the Real Frank Pantridge

## Frank Pantridge junior (Frank's nephew)

Frank's nephew is possibly the only person still alive who knew him best. He was a particular favourite of the great man as he was growing up. To avoid confusion I will use (j) after Frank junior and (s) after Frank senior in this section. Frank (j) told me:

> *His life was less ordinary, and he was less than ordinary himself. Indeed he was somewhat eccentric, although he was perfectly aware of this. He was simply a law unto himself. In my opinion he was paternalistic. He called me an intelligent idiot, so presumably I needed his tutelage. He told me about a junior doctor on his ward, who was mathematically gifted. The junior doctor suggested cutting a patient's artery as a way to address his dangerously high blood pressure. Uncle Frank said 'That was mathematically correct, but it would kill the patient.' I remember when he was dying in hospital: I said that I would be back to see him later in the day and he told me that, if I met a pretty girl, I was not to bother. You laugh, but you cry.*

Frank (j) told me that his uncle's lifestyle was pretty basic, and he was not motivated by material things:

> *He didn't want to own a yacht or a flash car. He didn't aspire to be a member of the jet set. He wasn't flashy. I guess after his experiences of war, he had a better appreciation of what was important in life. He wanted to be comfortable obviously. I remember he said to me*

*once that nobody needs more than £10,000 a year to live on. I don't know what that would be in today's money. So, he was a little left centre, although if you didn't know him you might think he was right wing. He seemed to be fond of the quote 'money is important when you don't have any.' He did however like the odd drink and I use that word loosely. He preferred malt whiskey, only ever single malt because he said that he did not know what was in blends. He only drank German beer because 'only the Germans know how to make good beer.'*

Frank (s) was very fond of his nephew and niece and Frank (j) recalls an amusing story during a visit to see him whilst they were young children:

*I remember when my sister and I would go and visit him as children (he used to complain that we never came to see Uncle Frank) and at some point, we would generally take a stroll outside weather permitting. We would all be walking and then something would occur to him and he would suddenly stop – and of course we would also stop. He would briefly consider whatever the point was and then we would move off again. I remember on one occasion he told us that he was going to write a book on what to say to yourself when you talk to yourself. Another day when we were visiting him, the postman arrived and knocked on the door. The postman was looking for a house number on Corcreeny Road that did not exist. Uncle Frank said to him 'Well, this is number twenty two, and over there, on the other side of the road, is number twenty three, and if you continue to the end of the road and turn right, you will come to a little church and if you go in there and say a prayer, maybe God will tell you where number twenty is.'*

Frank (j) became aware that his uncle, like my own father, would never forgive the Japanese for their mistreatment of prisoners of war. *'What good would an apology from the Japanese be to me?'* He knew that the Japanese would never pay him compensation at the level of £75,000, for each day that he spent in captivity – which was the figure he came up with 20 years ago. Ironically, in 1951, under the San Francisco peace treaty, former PoWs like Frank and my father were given £76.

Frank(j) tells me about an incident with a nurse:

*At my cousin Elizabeth's wedding, someone asked him why he never got married and he quipped that 'I would not marry anyone who would be mad enough to marry me.' He was certainly married to his work. It was sad that that he would be in pubs and restaurants, sitting there on his own, with no wife. He once asked a nurse called Mabel Stevenson out. She said yes. He picked her up in his car and enquired where she wanted to go. She replied that she didn't know and that she would leave the choice of restaurant to him. With that he turned the car and dropped her off at her house. 'There's really no point in going any further, I can't stand a woman who can't make up her mind, next time somebody asks you out, you will know where you want to go.' This story went round the hospital. I was sad that he was in pubs and restaurants, sitting there on his own, with no wife.*

He went on to tell me: *'My uncle used to talk about "Columbus syndrome". Columbus didn't know where he was going. When he got there, he didn't know where he was, and when he got back, he didn't know where he had been.'*

Frank (j) also recalls his uncle's posture and the way he would walk around:

*He adopted a military posture – you know: chin up, chest out, shoulders back and stomach in. He once met some children on the road and he pointed to his house and said, 'I wouldn't go near that house over there, if I were you; a very bad man lives there.' It was suggested to me by a colleague of his that he was actually shy, but I questioned whether Uncle Frank was actually shy, or whether he was creating a smokescreen as a way to keep people guessing. He would often talk away to waitresses in a restaurant. The colleague replied that 'yes, but he was able to control them.' If the police (whom he referred to as the 'Gestapo' but not in a hostile way) stopped his car and asked him where he was going, he would say 'I'm going home'. Of course, the police would then want to know where 'home' was, and he would say, 'home is where I live.'*

When Frank (j) visited him in the hospital when he was dying of heart failure, he recalls:

*He had these pictures of the graves of the men who died in the Far East sitting on a table. I think they came from Councillor Jim Dillon. I asked him if he knew any of them. His reply was something like, 'I'm sure I knew some of them.'*

*I remember uncle Frank falling over one night in his house. He had heart failure, which (he probably knew) would kill him in months. He managed to alert us and we got him into bed. Then he got out his wallet and was going to give me money. It was about one o'clock in the morning and I asked him if he would like me to stay with him for a while. He replied that it would be a good idea. In contrast, on the evening of his death in the RVH, his final words to me were "quiet, quiet" which was interesting, given the title of his autobiography.'*

Frank's friend and colleague John Geddes, the man who helped him develop the defibrillator was asked what he remembers most about him. His response was '*It was a jab in the chest.*'

'*He wrote out the words (which he took from a poem) "Let us be up and doing, the grave is not the goal" and told me to pin it to my bedroom wall.*'

Frank junior's account of his uncle provides an insight into the amazing man who endured the horrors of Japanese camps and returned to give society one of the most important inventions of the twentieth century.

## Lady Mary Elizabeth Peters LG CH DBE DStJ

Mary Peters is probably one of the best-known sports personalities in Northern Ireland, possibly only surpassed by the late George Best. She shot to fame in 1970 when she won two gold medals for Northern Ireland in the Commonwealth Games held in Edinburgh. Two years later 1972, she won the gold medal in the Pentathlon at the Munich Olympic games, the only track athlete to win a gold medal for Great Britain and Northern Ireland at the games. Her win over the home favourite Heide Rosendahl became a massive international story. She came from a Northern Ireland blighted by violence and her resilience was evident from her earliest days, travelling across battleground Belfast with her athletics kit.

On her return to Belfast after the games she, like Frank Pantridge, was subject to IRA death threats. A phone call to the BBC after she won the gold medal gave a chilling message:

*Mary Peters is a Protestant and has won a medal for Britain. An attempt will be made on her life and it will be blamed on the IRA… her home will be going up in the near future.*

Mary insisted on returning to Northern Ireland despite offers of jobs in the United States and Australia, where her father lived. On her return, she was feted as a heroine by the success-starved people of the province and paraded through the streets of Belfast. Due to the seriousness of the threat, she was not allowed to return to her flat for three months after her return.

Mary first met Frank Pantridge in a pub in Hillsborough called The Plough, his favourite watering hole not far from his home. She knew of him as she had been a patron of the Royal Victoria Hospital where he was a consultant. Mary takes up the story of this first meeting:

> *I saw him in the pub that night drinking with somebody else and I could sense them talking about me. Eventually he came over and shook my hand. I recognised that perhaps he had had a few too many. They used to lock his car in the car park so that he couldn't drive it home as he had to cross the main road to get home. Frank thought he owned the world, he was very strong willed and if he wanted to get from A to B he would just do it despite anybody else who might be inconvenienced.*

Mary became friendly with Frank later when she became a patron of the Friends of the Royal Victoria Hospital Trust.

In 2001 when Jim Dillon became Mayor of Lisburn, he had to choose a charity and as he knew Frank, he set up The Pantridge Trust which Mary was invited to join. This charity's aims were to put more defibrillators out into the workplace in Northern Ireland. Both Frank and Mary were by now both freemen of the city of Lisburn. Another six people were invited to join the trust: Dr Norman Campbell, Professor Alun Evans, Jack Kyle, Dr Charles Wilson and Frank's nephew, Frank Pantridge junior. Jack Kyle was particularly indebted to Frank, as he had saved his mother's life.

Mary recalls one particular Pantridge Trust charity event in the Lisburn Civic Centre:

> *We had our photographs taken with all the dignitaries and it was the installation of a new mayor. He saw a politician he didn't like too much, and he said to me, I'm going to tell him what I think of him'. I said, 'You will not, Frank Pantridge.' I took his arm and said, 'You will not lose your dignity, another time perhaps.'*

As already alluded to earlier in the book, Frank was totally intolerant of the Japanese or anything Japanese. Mary recalls Frank taking his car, not a Japanese one, into the garage to have a fault with the engine fixed; when he went to pick it up the mechanic said, 'The problem is these Japanese engines.' Frank refused to take the car out of the garage.

When Frank died, Mary was on holiday in the Isle of Wight with the late Cliff Morgan and his wife. Morgan was a former Welsh rugby union player and broadcaster. It wasn't until she came home that she heard of his death and that he had already been buried in the family grave in St Malachy's Church, Hillsborough.

Along with Jim Dillon and the other Pantridge Trust members, Mary arranged to have a statue of him made. Sculpted by John Sherlock, Pantridge's statue now sits proudly outside Lisburn Civic Centre main entrance. It was unveiled on 12 February 2006 at a memorial service, by Timothy Knatchbull, the grandson of Lord Louis Mountbatten. Knatchbull was in the boat along with his twin brother when it was blown up by an IRA bomb near Mullaghmore, a seaside village in County Sligo, just 12 miles from the Northern Irish border. The boat was destroyed by the blast, and Mountbatten was pulled alive from the boat, but died of his injuries later. Knatchbull's twin brother Nicholas was also killed in the blast.

Ironically, in 2018, Mary herself suffered heart problems and had to undergo six hours of open-heart surgery in the cardio department of the RVH where Frank had been a consultant before he retired. Without the surgery she had been told that her life expectancy could be as little as two years. She was fitted with a new heart valve and is now full of vitality and health, as she says herself, 'I feel alive again.'

During my interview with Mary, she backs up what many people have said about Frank Pantridge:

*He fell out with a lot of people in the hospital, he was intolerant of people who he thought were ignorant of his ability and of his knowledge, he didn't suffer fools gladly either. I loved him though,*

*I thought he was a real character. You either loved Frank or hated him, he was like marmite. I loved and respected him because I knew how many lives he saved. I found him intriguing as a person. He was always grumpy, but he loved people.*

## Jim Dillon – former Mayor of Lisburn and Castlereagh Council – 2000–2002

Jim Dillon first encountered Frank Pantridge on the roads around his farm near Moira. He recalls:

*I would often see him walking the roads and I would always give him a wave as I drove past. In 1981 I went on a business trip to California and to my great surprise they knew more about Pantridge than we did in Northern Ireland. His defibrillators were everywhere, and I met a personal friend of his in Los Angeles and he said, 'Why is something not done in Northern Ireland about Frank? That man should have a knighthood.'*

When Jim got back, he decided to call and see Frank and one evening knocked on his door. The two chatted for quite a while about the development of the defibrillator and about Frank's life during the war. Jim got the impression that he was bitter about the fact that his achievements had not been recognised fully and that he had not been knighted. He said that:

*Frank was proposed three time for a knighthood, but was turned down each time, he rubbed a lot of people up the wrong way – he would have just sent you to hell.*

Whilst Frank fell out with many people over the years, he never fell out with Jim Dillon. The two men formed a bond of friendship and as Jim says:

*He just loved my wife and every other Saturday night he would take us out for dinner. I think he liked my wife because, as she didn't drink, she would always drive the two of them to the pub.*

Jim recalls a story Frank told him about drinking in The Plough one evening:

*This young girl of about 19 came over to me and said, 'Are you Frank Pantridge?' 'Yes I am,' I said. 'I want to thank you profusely because I wouldn't have been in this world only for you,' she said. 'Oh, I don't recall being with your mother.' 'Oh nothing like that, you saved my mother's life twenty-one years ago.'*

Jim, along with Mary Peters, was the driving force behind the commissioning and erection of Frank's statue outside the Lisburn City Council offices.

*Appendix III*

# Famous People Whose Lives Have Been Saved by a Defibrillator

*Sudden cardiac arrest can strike anyone — very often its victims are people in middle age with no previous history of heart problems. Celebrities with access to top-rate preventive care, surrounded by staff and personal assistants are no exception when sudden cardiac arrest strikes.*

### Lyndon B. Johnson President of the United States of America 1963 – 1969

In April 1972 American President Lyndon B. Johnson had a heart attack whilst visiting his daughter in Charlottesville, Virginia. He was resuscitated by a mobile coronary care unit using an AED originally designed by Frank Pantridge and his life was saved. He credits Pantridge:

> *Almost certainly, only one cardiologist has conducted pioneering work that saved my life and those of thousands of others. This distinction goes to Dr James Francis "Frank" Pantridge, professor at Queen's University, Belfast, Northern Ireland. During a visit to Charlottesville, Virginia, I had a heart attack in 1972, which a mobile coronary care unit successfully treated with a Pantridge defibrillator. I owe my life to the invention of this former Japanese prisoner of war.*

## Fabrice Muamba – Former professional footballer

On 17 March 2012, forty-three minutes after kick-off in the FA Cup tie between Bolton Wanderers and Tottenham Hotspur, played at White Hart Lane, north London, home of Spurs, 23-year-old Fabrice Muamba collapsed. As the stadium fell silent and medical staff huddled around him, the match was abandoned. Fabrice's heart had stopped beating for seventy-eight minutes.

The first minutes after he collapsed were crucial. According to Amy Lawrence, who had been covering the match for *The Observer*:

> *The reason everyone knew something was wrong was that he, though out of the current run of play, suddenly 'fell like a tree trunk'. He didn't put his arms out to break his fall, or anything, he just dropped. It was seconds before other players noticed. Rafael van der Vaart, a Spurs player, was the first to do so, and frantic signalling to the pitch-side medical teams brought on the men in green.*

Spurs had five fully medically trained assistants pitch side that day, and there was, as ever, aid from the St John Ambulance unit. They had the defibrillator ready. Fabrice was given fifteen defibrillation shocks in all: two on the pitch, one in the tunnel and twelve in the ambulance. The heart needs to be jump-started with a 300-joule electric shock. In total, he was to take 4,500 joules in those seventy-eight minutes. As he entered the tunnel, the crowd finally fell silent. A further twelve defibs were carried out during the journey to hospital. Fabrice later recalls the incident and talks about the device that saved his life:

> *A person lying prostrate after a sudden cardiac arrest, surrounded by others frantically trying to revive them. This is a scene that became very real for me when I collapsed on the pitch during an FA cup quarter-final against Tottenham Hotspur on 17 March 2012.*

*The medical staff who treated me that night were incredible and I will forever be indebted to them. But I also owe my life to Professor Frank Pantridge, the man who invented the portable defibrillator; a device that has helped save millions of lives across the world over the past fifty years, a man who survived three and half years of brutal treatment in Japanese prisoner of war camps. March 17 2012 is a date that will be forever etched in my mind, as it was the day my life changed completely when my heart stopped for seventy-eight minutes. Frank Pantridge was a remarkable man. He died in 2004, but his contribution to cardiology lives on. Portable defibrillators are used throughout the world to save lives today and the process emergency departments still use to treat out-of-hospital cardiac arrest is referred to as the 'Belfast Protocol'.*

*The late professor John Anderson worked with Frank Pantridge in Belfast during the 1960s and 70s. He founded HeartSine Technologies in 1998, which manufactured defibrillators (also known as automated external defibrillators – AEDs). Every life saved by a HeartSine defibrillator is recorded by the firm. When data from the device is relayed around Northern Ireland, the people who made it are given a round of applause.*

*When someone goes into cardiac arrest, every minute without CPR and defibrillation reduces their chance of survival by ten per cent, so it is vital we continue to build on Frank Pantridge's legacy by having more portable defibrillators readily available.*

## Bernard Gallacher OBE – former Ryder Cup Golfer

Born in Bathgate, Scotland, Bernard Gallacher took up golf at the age of eleven. In 1965 he won the Lothians Golf Association Boys Championship and two years later won the Scottish Amateur Open Stroke Play Championship, before turning professional. Gallacher accumulated ten wins on the European Tour between 1974 and 1984 and finished in the top ten on the European Tour Order of Merit five times between 1972 and 1982, with a best placing of third in 1974. He went on to play in the Ryder Cup eight times and was non-playing captain of the European Team in 1991, 1993 and 1995.

On 29 August 2012, thanks to the availability of an automated external defibrillator following his collapse during a dinner reception in Aberdeen, along with early-response CPR treatment at the venue, Bernard Gallacher's life was saved:

*I consider myself incredibly fortunate that a defibrillator was on hand in Aberdeen to resuscitate me. But for that AED and the quick-thinking, expertise and life-saving skills of the medical personnel in the room that night, I wouldn't be here today.*

*'I died, medically speaking, three times,' he said, as he explained why he has grown used to thinking about mortality and the mysteries of an after-life.*

*I had a cardiac arrest rather than a heart attack – which is associated with the plumbing of the heart, like a coronary with blocked arteries. I was just about to stand up and talk at a golf function in Aberdeen. People tell me that I fell over like I was hit by a boxer. I was poleaxed. I was just about to say something and ...'*

An unlikely chain of coincidences rescued him – the first being that he was attended to immediately by an experienced A&E nurse:

*She gave me mouth-to-mouth for 20 minutes before they got the defibrillator going. Apparently, I played golf earlier that day with this guy who teaches people how to use defibrillators as part of his health and safety work. He got the hotel defibrillator working.*

During a fraught trip in the ambulance to hospital, Gallacher's heart stopped completely again and again:

*They had to use the defibrillator three times. A South African, Doctor Mark Bloch, saved me three times in the ambulance. In*

> terms of the heart not working, I really was gone. I was medically dead. But he kept me going. I was very lucky. Around 100,000 people in the UK suffer from cardiac arrest every year. And if there is a defibrillator around you have over a twenty per cent chance of survival. If not, your chances reduce to around five per cent. Those statistics are frightening – but they inspired my wife. She's the driving force behind the campaign. We're going to run with it until the Ryder Cup in September and Lesley will remain involved with the Arrhythmia Alliance.

Gallacher now has an ICD implanted in his chest as he is at risk of further heart problems.

## Glenn Hoddle – ex-England International footballer and former England manager

Sixty-two-year old Glen Hoddle is another celebrity who was to pay thanks to Frank Pantridge's invention. Whilst working as a football pundit for BT Sport in their London studios, he collapsed and hit his head on the floor. Fortunately, a technician called Simon Daniels was in the studio and he immediately grabbed a defibrillator. After Daniel's hard work, which saved Hoddle's life, he was taken away in an air ambulance. Hoddle had no memory of all of this and when he came out of hospital, his colleagues including Robbie Savage filled him in with how events unfolded that day.

> The only other thing I remember is my son Jamie, who lives in London, he was one of the first to arrive and I remember just squeezing his hand and seeing him in the ambulance. My heart stopped for at least sixty seconds, I was gone. The engineer ran to me and brought me back to life using the defibrillator. I wouldn't be here if not for him and the fact that defibrillators are widely available now. It just wasn't my time to go.

Following the incident, Hoddle has teamed up with the British Heart Foundation, and he admitted his delight at a Government ruling that means CPR and defibrillator use must be taught in schools, starting next year. 'I was very happy to hear that news. What's more important than saving a life? I'm so glad that the Government has decided to look at this with a view to adding it to the curriculum.'

# Bibliography

Abbott S., *And All My War Is Done*, Pentland, Durham, 1992
Arnold M., *The Sacrifice of Singapore,* Marshall Cavendish, Singapore, 2001
Barber N., *Sinister Twilight*, Cassell, London, 1968
Barnard J., *The Endless Years,* Chantry, London, 1950
Barwick I., *In the Shadow of Death,* Pen & Sword, 2003
Bayley C. & Harper T., *Forgotten Armies,* Penguin, London, 2005
Beattie R., *The Death Railway*, TBRC, Kanchanaburi, 2015
Braddon R., *The Naked Island,* Birlinn, Edinburgh, 2005
Bradley J., *Towards the Setting Sun,* Fuller, Wellington, 1982
Bruton P., *British Military Hospital Singapore*, Bruton
Churchill W., *The Second World War*, Pimlico, London, 2002
Clarke H., *A life for Every Sleeper*, Allen and Unwin, Sydney, 1986
Coogan A., *Tomorrow You Die*, Mainstream, Edinburgh, 2013
Cosford J., *Line of Lost Lives*, Gryphon, Northants, 1988
Davies P., *The Man Behind the Bridge*, Athlone, London, 1991
Daws G., *Prisoners of the Japanese,* Robson, London, 1995
Elphick P., *Singapore: The Pregnable Fortress*, Coronet Books, London, 1995
Farrell B., *The Defence and Fall of Singapore*, Tempus, Stroud, 2005
Hack K. & Blackburn K., *Did Singapore Have to Fall*, Routledge, London, 2003
Hastain R., *White Coolie*, Hodder and Stoughton, London, 1947
Hastings M., *Nemesis*, Harper, London, 2007
Hayes D., Noble M., Fallon R., *Understanding Your Pacemaker or Defibrillator*, 2012
Horner R., *Singapore Diary*, Spellmount, Stroud, 2007
Holmes R. & Kemp A., *The Bitter End*, Bird Publications, Chichester, 1982
Kennedy J., *Andy Dillon's Ill-Fated Division,* United, Cornwall, 2000
Lane, A., *Seventy Days to Hell*, Lane Publishing, Stockport, 2011
Lane A., *When You Go Home*, Lane Publishing, Stockport, 1993
Lane A., *Lesser Gods Greater Devils*, Lane, Stockport, 1993
Lomax E., *The Railway Man*, Cape, London, 1995
Low, N. & Cheng H., *This Singapore, our city of dreadful night*, Singapore, 1945
Lowry C., *Last Post over the River Kwai*, Pen & Sword, Barnsley, 2017

Mitchell S., *Scattered Under the Rising Sun*, Pen & Sword, Barnsley, 2012

McEwan A. & Thomson C., *Death was our Bedmate*, Pen & Sword, Barnsley, 2013

McEwan J., *Out of the Depths of Hell*, Pen & Sword, Barnsley, 2005

McArthur B., *Surviving the Sword*, Time Warner, London, 2005

Moffatt J. & Holmes A., *Moon Over Malaya*, Tempus, Stroud, 2002

Partridge J., *Alexandra Hospital Singapore*, Singapore Polytechnic, 1998

Pantridge F., *An Unquiet Life*, W & G Baird Ltd, Belfast, 1989

Peek I., *One Fourteenth of an Elephant*, Transworld, London, 2004

Russell, Edward Lord, *Knights of Bushido*, Corgi, London, 1960

Searle R., *To the Kwai and Back*, Collins, London, 1986

Smith C., *Singapore Burning*, Viking, London, 2005

Summers J., *The Colonel of Tamarkan*, Simon and Schuster, London, 2005

Taylor E., *Faith, Hope and Rice*, Pen & Sword, Barnsley, 2015

Tsuji M., *Japan's Greatest Victory, Britain's Worst Defeat*, Spellmount, Kent, 1997

Thompson P., *The Battle for Singapore*, Portrait, London, 2006

Woodburn Kirby, *The War Against Japan*, Her Majesty's Stationary Office, London, 1968

Wyatt J. & Lowry C., *No Mercy from the Japanese*, Pen & Sword, Barnsley, 2008

# Index

A Force, 57
Alexandra Hospital, Singapore 28–9, 34
Allison, R.S., 95
*Almanzora*, SS, 89–91
Alor Setar, xvii
American Heart Association, 115
Anderson, Doctor John, 102
Annandan, Private A., 30
Apthorpe, Captain D., 57
Argyll & Sutherland Highlanders, 24
Australian forces, 23, 28
Automatic Implantable Cardioverter-defibrillator, 113–14
*Automedan*, SS, 37

Ban Pong, 57, 61–2, 66
Bangkok, 57
Barnard, Professor Christiaan, 105
Beattie, Rod, xii, 57
Beckworth-Smith, Major General M., 26
Belfast, 1–7, 10, 14, 47, 53, 93–5, 97–8, 100–103, 105, 111, 123, 128–9, 143
Bencoolin Street, Singapore, 9–10
Beriberi, 73, 95, 127
Best, George, 121, 138
Biggart, Sir John, 109
Bingham, John, 98
Birdwood Camp, Singapore, 24–5
Blair, Tony, 117
Boyd Campbell, Doctor, 4
Brabourne, Baroness, 106, 108, 120
Brackman, Arnold C., 116
Brewster, Sergeant W, 23
British Heart Foundation, 101, 129

British Legion, 17
British Medical Journal, 133
Brooke Popham, Air Chief Marshall HRM, 11–12, 18, 20
Brown, Tina, 105
Bull, Major, 29
Bull, Professor John, 99, 131–2
Burma, 14, 56, 73–6, 80, 127
Bushido, 86
Bye, Doctor W., 81

Canadian Heart Foundation, 104
Capper, D., BBC, 107
Cardiopulmonary resuscitation, xv
*Celebes Maru*, 57
Chamberlain, Neville, 6
Changi, prisoner of war camp, xiv, xvii, 8, 13, 40–2, 45, 47–9, 52–8, 68, 78, 81–2, 88, 110, 127
China, 17
Churchill, Sir Winston, xiii, 7, 11–12, 28, 34, 36–7, 94, 109
City Hospital, Belfast, 4
Clarke, Sir Kenneth, 112
Columbo, Ceylon, 91
Commonwealth Graves Commission, 78
Craven, Lieutenant Colonel J.W., 12, 29
Crocker, General Sir J.T., 92

*Daily Telegraph*, 91
Daley, F., 98
DHSS, 109–11
Dillon, J., 6, 139–42
Dillon, Lieutenant Colonel A., 72
Downpatrick, xiii, 1, 23

Downshire School, 1–2
*Duchess of Bedford*, SS, 23–4
Duckworth, Padre J.N., 60
Duff-Cooper, Sir A.D., 11
Dunbar, Sir R., 122

East Surrey Regiment, xiii, xvii, 16, 21, 97
Eighteenth Division, 22, 25
*Empress of Australia*, RMS, 92
European Congress of Cardiology, 130
Evans, Professor A., 6, 10, 123, 125, 133, 139
Evans, R., 11

F Force, 14, 60–1, 64–7, 69, 72–3, 78
Field, Doctor T., 10, 126–7
Fitzgerald, G., 1
Flanagan, Sir J., 108–109
Fort Caning, Singapore, 38
Friends School, Lisburn, 2, 126
Fukuei, Lieutenant General, 52

Gallagher, B., 145–7
Galleghan, Lieutenant Colonel, 21
Geddes, Doctor J., 102,129
Gemas, Malaysia, 21
Geneva Convention, 43, 53
Gibraltar, 91
Gordon Highlanders, xiii–xiv, 13–17, 21–8, 34–5, 40–6, 126
Graham, Lieutenant Colonel, 13
Grigg, Sir J., 80
Gurun, battle of, 97

Harewood, Lord H.E., 120
Harland & Wolfe, 4, 90
Harris, Lieutenant Colonel, 61
Hawke, Prime Minister B.,116
Heath, Prime Minister E., 104
Hillsborough, xiv, 1–3, 6, 104, 121, 126, 138, 140
Hirohito, Emperor, 82, 87, 107, 116
Hirshoma, 13, 82, 88, 96, 118
Hoddle, G., 147–8

Holmes, Lieutenant Colonel W.G., 47, 51–2
Hunt, Major B., 66, 73

Imperial Japanese Army, 10, 45
Imperial War Museum, 107, 120
Indian Brigade, 11th, xvii, 16, 20–1
3rd Indian Corps, 23
4th Indian Brigade, 29
Irish Republican Army, 1, 104–105, 107–108, 138, 140

Japanese Imperial Guards, 22, 27
Jaywick, Operation, 15
Jitra, Battle of, 16, 20–1
Johnson, Doctor S., 106–107
Johnson, President L.B., 9, 103, 143
Johore, 11, 16, 22–3, 25–6
Jones, Captain, 47

Kanchanaburi, 75–7
Kempeitai, 55
Keppel Harbour, Singapore, 23, 25, 87–9, 91
Knatchbull, T&N, 106, 140
Konkoita, 75–6
Kouwenhoven, W.B., xv
Kwai, river, 57, 77
Kyle, Jack, 123, 139

Lancet, 97, 101, 103
Lane, A., 48–9
Lisburn, 6–7, 15, 115, 121, 132, 139–40, 142
Lomax, E., 85
Londonderry, 115
Long Kesh, 106
Lown, B., 113
Lyon, Lieutenant I., 15, 23

Magheralin, 1–2
Malay Volunteer Force, 93
Malaya, xiii, 1, 7–12, 16–19
Manchester Regiment, 48

Manchuria, 13
Marshall, Doctor R., 97
Mass, Doctor F., 15
Mawhinney, A., 100
Maxwell, P., 106
Mayrs, Professor, 108
McArthur, General D., 87
McArthur, Sir W., 55
McNeilly, H., 99, 128
Michigan, University, 96
Military Cross, 10, 26
Milliken Doctor T., 88–9, 127
Mitral Stenosis, 98
Moore, Lieutenant, 34
Moulmein, 57
Mountbatten, Lord Louis, 87, 105–107, 131, 140
Mower, MM, 113
Muanga, F, 144–15

Nagasaki, 13, 82, 96, 118
NASA, 102
Needham, R., 112
New York Times, 110
NHS, 96, 113
Non Pladuk, 76

Oakley, Doctor C., 129
Okami, Lieutenant Colonel, 50,52
Okusalii, Lieutenant, 48–9
*Ormonde*, SS, 92

Paisley, Sir I., 1, 21
*Pampanito*, USS, 79
Pantridge Trust, xi, 139
Pantridge, F. Jnr, 6, 134–7, 139
Pemberton, J., 99, 128
Penerang, Malaysia, 16, 21, 43
Percival, General A., 10, 19, 23, 35, 42
Peters, Lady M., 6, 121, 138–40, 142
Phillips, Admiral T., 20
Pond, Lieutenant Colonel, 72
*Prince of Wales*, HMS, 20, 126
Prince Philip, 105

Queen Elizabeth II, 105, 107
Queens University Belfast, xiv, 3, 5, 55, 88-9, 95, 99, 105, 108, 123, 126, 132

RAAF, 18, 85
Raffles Hotel, 86
*Rakuyo Maru*, 79
Rangoon, Burma, 56–7
RAPWI, 85
Red Cross, 54, 74–5
*Repulse*, HMS, 20, 126
Roberts Hospital, 78
Royal Air Force, 11–12, 18, 37, 85, 106
Royal Army Medical Corps, 6, 14, 126
Royal College of Physicians, 96
Royal County Down, 1
Royal Norfolk Regiment, 57
Royal Ulster Constabulary, 108
Royal Victoria Hospital, x, xvi, 4–6, 53, 94–9, 104, 106, 108, 121, 123, 127–8, 132, 138–40

Sands, B., 121
Searle, R., 107
Seattle, 103, 111, 129
Selerang, Singapore, 13–14, 17, 23, 41, 46, 50, 52–3, 79, 81
Shanghai, xvii
Shenton-Thomas, Sir T., 12, 19, 35
Sherlock, J., 121, 140
Shimpei, General, 46
Sime Road POW camp, 90
Simkin, Lance Corporal, 30
Simpson, Brigadier I., 11, 23, 35
Sligo, 105–106, 140
Smiley, Captain, 28, 31
Smith, Padre, 30
Songkurai, 60, 63, 66, 68, 70–2
Southampton, 91, 93-4
Spiers, J., 93
Staunton, W., 108
Stitt, Lieutenant Colonel, 18, 43
Stormont, 104
Straits Times, 90

*Strathmore*, SS, 8, 87
Suez Canal, 8, 91
*Sussex*, HMS, 88

Takanum Camp, 72
Tenga Aerodrome, 27
Thanbaya, 12, 72–3, 76, 127
Thanbyuzayat, 57
Thatcher, Baroness, 1
Thompson, Captain H.B., 97
Thompson, Sir W., 97
Three Pagoda Pass, 70
Time Magazine, 102
Times, 111
Toosey, Sir Philip, Lieutenant Colonel, 77
*Toyohashi Maru*, 57

Trincomalee, 91
Truman, President, 13, 96, 119
Tsuji, Colonel, 119

Ulster Division, 2
Ulster Transport Museum, 7
Uruguay, 114

Ventricular fibrillation, 14–15, 99–100

Wampo Viaduct, 77
Wavell, General Sir A., 23–4, 27–8, 34, 123
Wedge, M., 123
Whitelaw, Sir W., 104–105
Wild, Colonel CDH, 49
William of Orange, 1
Wilson, Doctor C., 7, 101, 130, 139
Wilson, F.N., 128
WRENS, 91

Yamashita, Lieutenant General, 38–9

Zipper, Operation, 87